MECHATRONICS – THE INTEGRATION OF ENGINEERING DESIGN

Papers prepared for the University of Dundee
and the Solid Mechanics and Machine Systems Group of the
Institution of Mechanical Engineers

Published by
Mechanical Engineering Publications Limited
LONDON

© The Institution of Mechanical Engineers 1992

This publication is copyright under the Berne Convention and the International Copyright Convention. All rights reserved. Apart from any fair dealing for the purpose of private study, research, criticism or review, as permitted under the Copyright, Designs and Patents Act, 1988, no part of this publication may be reproduced, stored in a retrieval system, or transmitted in any form or by any means, without the prior permission of the copyright owners. Reprographic reproduction is permitted only in accordance with the terms of licences issued by the Copyright licensing Agency, 90 Tottenham Court Road, London W1P 9HE. Unlicensed multiple copying of the contents of the publication without permission is illegal. Inquiries should be addressed to: The Managing Editor, Mechanical Engineering Publications Limited.

Authorization to photocopy items for personal or internal use, or the internal or personal use of specific clients, is granted by the Institution of Mechanical Engineers for libraries and other users registered with the Copyright Clearance Center (CCC), provided that the base fee of $3.00 per paper plus $0.05 per page is paid direct to CCC, 21 Congress Street, Salem, Ma 01970, USA. This authorization does not extend to other kinds of copying, such as copying for general distribution, for advertising or promotional purposes, for creating new collective works, or for resale. No copying fees are payable for papers published prior to 1978.
0957-6509/91 $3.00 + .05

The publishers are not responsible for any statement made in this publication. Data, discussion and conclusions developed by authors are for information only and are not intended for use without independent substantiating investigation on the part of potential users.

ISBN 0 85298 840 0

A CIP catalogue record for this book is available from
the British Library

Printed by Moreton Hall Press Ltd,
Bury St. Edmunds, Suffolk

CONTENTS

Schemebuilder and Layout: computer-based tools to aid the design of mechatronic systems 1
   R.H. Bracewell, R.V. Chaplin, and D.A. Bradley

A knowledge based approach to feature recognition in machining processes 7
   B.S. Falay and R.M. Parkin

Design of mechatronic systems using bond graphs 15
   P.J. Gawthrop

Design function deployment, a platform for cross-disciplinary product development 23
   A. Jebb, R.C. Edney, S. Sivaloganathan, and N.F.O. Evbuomwan

Development of an educational kit for mechatronics design 29
   G.R. Kiss and G.S. Bellis

Development of a new diagnostic system by Ford Motor Company and GenRad 37
   R.J. Shorter

From woven bags to expert systems and broken digits 41
   C.J. Fraser, J.S. Milne, and G.M. Logan

Integrating the cutting and sewing room of conventional garment manufacture through automated stripping 47
   C.A. Czarnecki, A. Paterson, and B. Bramer

Sensor integration in high speed packaging machinery 53
   E.J. Rushforth, T.C. West, and P.D. King

Modelling and control of an active actuator system: slow and fast modal effects 61
   Shengbao Li and R. Barron

The design of mechanical amplifiers using piezoelectric multilayer devices for use as fast actuators 69
   J.K. Thornley, T.G. King, and M.E. Preston

Accelerometers for suspension control 75
   P. Kellet

A rapid response thermal tactile sensor 81
   G.J. Monkman

A smart sensor for precision position measurement 87
   C. Butler and Q. Yang

Intelligent transducers for materials physical properties measurement 91
   G. Kulvietis and A. Daugela

The use of giant magnetostrictive materials in fast-acting actuators 95
   R.D. Adams, C.A. McMahon, and R. Thomas

Flexible joint control of a KUKA IR 161/60 industrial robot 101
   J. Swevers, D. Torfs, M. Adams, J. De Schutter, and H. Van Bussel

Integrated process and control design for fast coordinate measuring machine 109
   M.R. Katebi, J. McIntyre, T. Lee, and M.J. Grimble

Camshaft profile measurement – a mechatronic approach 117
   M.R. Jackson

The appropriate control of manipulators for rehabilitation robots 123
   P.J. Kyberd

Model reference adaptive control of a modular robot 131
   P.C. Mulders, A.P.M.A. Martens, and J. Jansen

Digital control of pump-actuated hydraulic manipulators 137
   J.E. Holt and D.E.B. Palmer

Active control of surge in a working gas turbine 145
   M. Harper, D.J. Allwright, and J.E.ffowcs Williams

Intelligent control of cone crushers 151
   R.A. Bearman and R.M. Parkin

An integrated mechatronic research cell for the decoration and assembly of scale models 159
   L.G. Trabasso, A.P. Slade, and J.R. Hewit

The Lancaster University Computerised Intelligent Excavator programme 163
   D.A. Bradley and D.W. Seward

Experience of mechatronics applied to the design and management of oil-hydraulic and pneumatic systems 169
   S. Ferraris, A. Lucifredi, and E. Ravina

Machine intelligence: the key capability of mechatronic products 175
   G. Rzevski

Mechatronics – meeting the technicians needs into the 21st Century 183
   C.C.B. Day

Schemebuilder and Layout: computer-based tools to aid the design of mechatronic systems

R H BRACEWELL, PhD, R V CHAPLIN, PhD, D A BRADLEY, PhD
Lancaster University, UK

A design tool, currently under development at Lancaster University, is described. This is a software package aimed at the conceptual stage of mechatronic product design and builds on Michael French's earlier work on systematic design. The software runs successfully in prototype form with a limited set of components, the output being in the form of a bond-graph-like connectivity diagram, component specifications and solid model drawings of the product layout.

1. Introduction

Lancaster University Engineering Design Centre (EDC) was set up in September 1990 to develop design aids useful in the field of conceptual design, particularly that of mechatronic products. In the original research proposal French et al.[1] referred to four main projects, Schemebuilder, Layout, Creativity Aids/Design Principles and Function Costing. Figure 1 is a diagram representing the relationships between these individual projects, the outputs of which are intended to be useful whether used separately or in concert. This paper describes the progress of the Schemebuilder and Layout projects. These have now been combined but they are best initially described separately.

2. Schemebuilder

Schemebuilder is intended to support a truly *mechatronic*[2] approach to the conceptual design of products that use a combination of mechanical, electronic and software components. The central point of this approach, is the need to enable the designer to allocate required functions to solution elements, that have been chosen freely and without bias from any of the aforementioned engineering domains. For example it is often desirable to re-allocate functions that traditionally have been performed mechanically or by analogue electronics, to digital-electronic or software solutions. This requires a 'technology-neutral' approach to the design process, concentrating firstly on defining the essential flows of information within the system.[3] 'Information' is here interpreted in the widest sense of the word to embrace communication of electrical and mechanical variables, including power flows, as well as its narrower 'Information Technology' meaning.

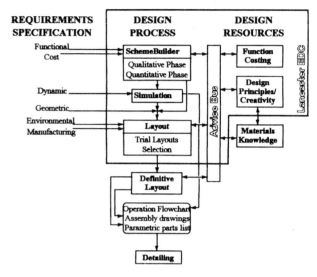

Fig 1   Relationships between Lancaster EDC projects

Buur and Andreasen[4] have argued that to represent mechatronic systems, new forms of design model are necessary, if this greater integration is to be achieved. Essential characteristics of these models should be as follows:

- They must provide a unifying notation, flexible enough to be applied and understood in all the diverse areas of mechatronics and forming a common medium for communication between specialists.

- They must be sufficiently comprehensive in scope to represent adequately any conceptual scheme that might be proposed to meet a particular specification.

- They must be expressive enough to allow multiple schemes to be evaluated and compared, exploiting fully the ever-growing analytical power of computers and knowledge-based software, so that the best may be reliably identified and selected for further development.

The major goal of the Schemebuilder project is to develop and demonstrate a system that can be used to rapidly produce, manipulate and evaluate such models.

2.1. Aims of the Schemebuilder Project

Accordingly, the objectives of the Schemebuilder project may be summarised thus:

- It should develop, in prototype form, a 'design workbench' software package for the conceptual design of mechatronic products. This is expressly *not* intended to be a fully automatic 'expert system designer', but a tool aimed at increasing the productivity and quality of output of human designers.

- The package will need to be able to hold a structured specification of the functions required of the product under development, allowing hardware or software components to mapped onto these functions, selected freely from the complete range of mechatronic technologies.

- The package should support the rapid creation, recording and evaluation of multiple competing solution schemes, proposed by the designer to meet the specification.

- It should provide a source of reference on technological areas unfamiliar to the user, giving as much encouragement as possible to explore solutions involving them.

- It is desirable that as much of this information as possible is stored in such a manner that the computer is able to reason about it as well as the designer. In this way the computer should be able to make informed suggestions as well as checking the designer's work for correctness and consistency.

2.2. Software Selection

A crucial early decision was the choice of software development package with which to create the Schemebuilder system. It was to run on one of 4 networked Sun Sparc 1+ UNIX workstations owned by the EDC. There were a number of criteria that the development software needed to fulfil:

- It would need to be suitable for a 'rapid prototyping' approach to software development, as much of the required functionality could not be precisely defined beforehand. A great deal of 'trying out ideas' would therefore be necessary. This requires very high-level programming tools to avoid much time and effort being wasted on work that might later be discarded.

- It would be important to produce clean, modular, extensible software and data structures to minimise code rewriting as the system expands and limit the scope of side effects caused by belated modifications. This and the fact that much of the mechatronic knowledge to be encoded in the system can be efficiently represented in terms of hierarchical structures, points strongly to an object-oriented approach.

- It should support modern graphical user interface (GUI) techniques such as windows, menus, icons and pointers (WIMPs) for greater ease of use of the finished package.

- It should support a full range of knowledge-based system development tools such as model-based reasoning, forward and backward rule chaining, demons, multiple simultaneous knowledge states and truth maintenance.

After consultation with the Artificial Intelligence Applications Institute (AIAI) in Edinburgh, the Common Lisp-based KEE software development environment from Intellicorp[5] was chosen as fulfilling all of these requirements. Experiences with using KEE have so far proved very positive.

A referee of this paper expressed surprise at the choice of KEE, making the valid point that accessibility to end users might be reduced by the additional expense of having to purchase a runtime KEE license. This this indeed a disadvantage, however it was (and still is!) felt that in view of strictly limited manpower, the overriding priority had to be obtaining the most powerful *development* tool available. In particular, the KEEWorlds and ATMS facilities are essential to further progress, but simply not available in current C or C++ based alternatives. If and when the prototypes are proven, the decision will have to be taken whether to recode or deliver in KEE.

2.3. Software Development

The initial strategy has been to create a library of mechatronic component classes, represented graphically to the user, instances of which can be created as required and joined together to form schemes. The component representations are inspired by Paynter's bond-graphs[6] - components have 'ports' by which they can be linked together and through which signals and power are communicated. Two compatible power ports linked together form a 'bond' and two signal ports form an 'active bond' in Paynter's terminology. As an example, figure 2 depicts a simple system along with its conventional bond-graph and Schemebuilder representations.

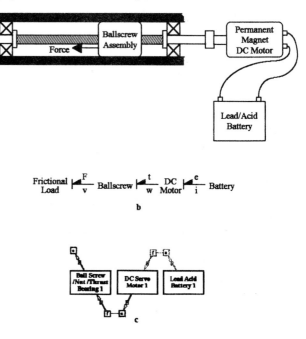

Fig 2 Simple system, bond graph, and schemebuilder representation

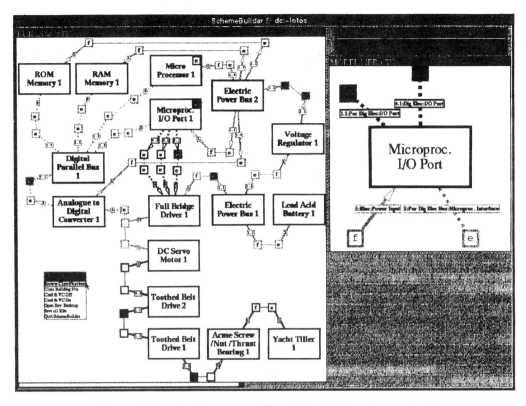

Fig 3 Schemebuilder screen with yacht autopilot under construction

Figure 3 is a screen dump of the current Schemebuilder user interface. The two prominent windows are the model library, in which the class of generic microprocessor I/O ports is currently displayed and the building site. The latter contains a number of instantiated components that have been inserted and joined together to form part of a scheme for a Yacht Autopilot. The component models stored in the library are accessed via a pop-up window, which allows browsing in any of three ways:

- by a conventional component classification, based on that of a commercially available product index
- by a classification of types port possessed by components
- a simple, alphabetical component list

From the browser, the user can choose either to display a component class in the model library or to add an instance of it to the building site at a specified location.

Figure 4 shows, in much simplified form, the distribution of information among and relationships between the six KEE knowledge-bases into which the current Schemebuilder software is modularised. Figure 5 shows the hierarchy of subclass links between units (objects) in the port knowledge-base: one of the three classification methods accessed by the browser.

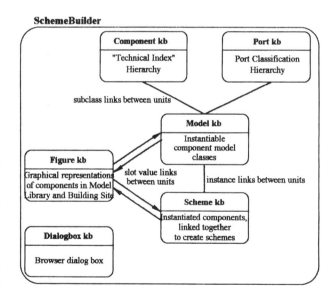

Fig 4 Schemebuilder knowledge-base (kb) relationships

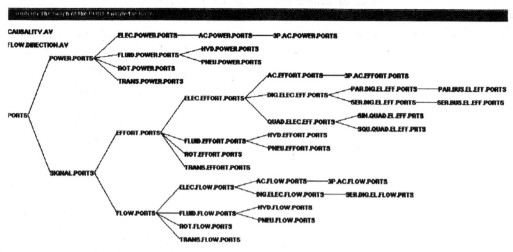

Fig 5 Hierarchical structure of the port knowledge-base

3. Layout

The Layout project supports the 'Embodiment of Schemes' phase of design. As the intention was to produce quick 'rough and ready' sketches of complete designs, it was thought[1] that the catalogue approach, as used by Ward and Seering[7] would load the system with too much data. Accordingly, it was decided that catalogue ranges of components would be described by drawings with dimensions given by parametric formulae. Such conventional parametric methods are described for example by Newell and Parden[8] and are used by component vendors, such as Festo[9] in their catalogues both on disc and paper.

3.1. Aim of the Layout Project

Briefly, the aim of this project is to assist the designer in arranging simplified representations of components in 3 dimensions, in order to satisfy the requirements set out in the product specification and the needs of manufacture, servicing and ergonomics. To this end, like Schemebuilder, it should enable the rapid creation, recording and comparison of multiple alternative layouts.

3.2. Parametrics

As the intention was to 'reserve spaces' in a design for particular components, these space-reserving drawings could afford to be much simpler than those of the real components, that would eventually replace them. For example, Figure 6 shows the parametric drawing used for stepper motors. It is of course a simplification of a catalogue drawing with features such as fixing holes, spigots and fillets omitted. A library of such drawings is being built up by extracting data from catalogues and suitably simplifying it.

Having made the simplified drawing, the dimensions of a particular component type can be related to its principal specifying parameter(s) by curve fitting, as shown in figure 7. Thus given, for example, the required output torque of a stepper motor, its space requirement can be sketched sufficiently accurately for layout purposes, using the set of curve-fit parametric formulae and the appropriate drawing.

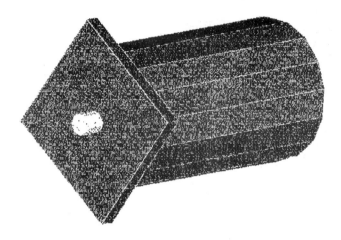

Fig 6 Parametric drawing of a stepper motor

In practice, these component drawings have been produced using a CAD macro language, the 'C'-based AutoCAD ADS/API. They are drawn as extruded or constructive solid geometry (CSG) solid models, allowing subsequent testing of the layout using sectioning and clash detection facilities provided by the CAD system.

As most real components are manufactured in discrete sizes within a range, the use of exact curves such as figure 7 would lead to difficulties at the detailing stage when the 'next larger' component would have to be selected. To overcome this problem we have used a Sheppard's correction,[10] based on the average step size in the component range:

$$correction = KS$$
$$\text{and } S = P_{n+1} - P_n \quad \quad 1$$

Where S is the step size, P_n is the specifying parameter value of the nth component in the range and K is a user-selected fraction of the step size, typically 0.5.

The procedure adopted in sizing a component is to add the correction(s) to the principal specifying parameter(s) before calculating the dimensions using the curve-fit parametric formulae. Figure 8 shows the resulting effect. Thus in laying out a number of such components it is likely that the overall space requirement will be similar in going from layout to detailing. Only minor changes in component positions should be required, as real components turn out to be a little larger or smaller than the parametrically sized ones.

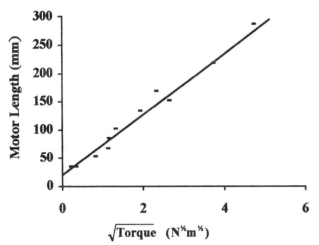

Fig 7   Curve Fitting

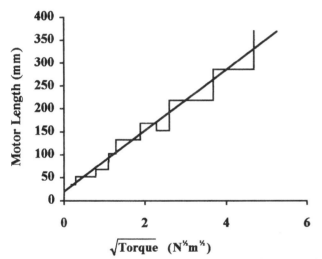

Fig 8   Sheppard's correction applied to a parametric formula

3.3. Software Selection and Communications

CAD Software

A number of CAD systems were considered for use, each offering solid model drawing within a macro language. These included Geomod, CIM-LINK, Personal Designer and AutoCAD Release 11. As it is a minor aim of the EDC as a whole to minimise software cost and to be accessible to a large group of users, AutoCAD was chosen and in particular the Advanced Modelling Extension (AME) solid modeller that has been developed from the earlier AutoSolid package.

Database software

Whereas the parametric formulae for dimensioning each component could have been accessed from a variety of database or general-purpose software development packages, the KEE environment was again chosen to guarantee good communications with Schemebuilder and allow an object-oriented treatment of the data.

Parametric formulae were written in LISP and linked to 'method slots' in KEE units (objects). Each component also has a slot containing the name of an AutoCAD external function, defined to draw it. Another method then uses the parametric formulae to calculate the component dimensions and via a UNIX named pipe, instruct AutoCAD to draw it.

Note that currently we run KEE and AutoCAD on separate SUN workstations and communicate between them. Software is however available that would allow them to run in overlapping windows on the same machine. While the UNIX pipe is set up communication is transparent and AutoCAD acts purely as a slave to the KEE process.

4. Fusion of the Schemebuilder and Layout Tools

Initially we had expected to pass the essentials of a scheme between the two tools as:

- a list of component names
- a list of corresponding parameters
- a list showing the functional links between components

However, as all this information for a particular design already exists in the scheme diagram built up on the building site, it seemed sensible to use this diagram to point directly to components in order to add them to an assembly drawing (as a check, building site components would be made to visibly flag that they had been added to the drawing).

4.1. Functions provided by the combined tool

Accordingly, a menu function was attached to each instantiated component drawn in the building site that offers the following functions:

Dimension.yourself	Calculate the component's dimensions
Draw.yourself	Draw the sized component at the origin
Weigh.yourself Size.yourself Cost.yourself	Calculate the Weight, Volume and Cost of the component and store these values in slots in the instantiated KEE unit
Position.yourself	This facility (to be added) will involve a control panel in KEE which should allow interactive positioning of each component in 3-D to form a Layout

The information channel between KEE and AutoCAD will carry, in one direction, a structured list comprising component name, dimensions and position co-ordinates and in the other direction the CAD handle for the solid model representing that component. The combined tool could similarly operate with other CAD packages via their macro languages using their solid modellers as 'drawing engines.'

4.2. Links with other projects in the EDC

Creativity/Principles

This project aims to prompt good design and stimulate

creativity by signposting good principles and/or by asking the designer to criticise the design on the basis of a set of design principles and so force creativity in attempting to fulfil their demands.[11]

The designer would refer to or be directed to a separate 'Index of Principles' as the product design was built up so optimising the design as it grew.

Function Costing

This project interacts with the design process at two stages:

- the conceptual stage
- the layout (or embodiment) stage

At the conceptual stage, 'true function costing' sets, for example, an envelope of least cost methods of providing a function, such as linear displacement, over the whole gamut of competing technical solutions and over the combined ranges of their principal specifying parameters. Such 'contour maps' of function cost would be used in guiding the creation of the initial function structure of the product, before Schemebuilder substituted alternative 'real' components into a scheme.

At the Layout stage, function cost would provide well-founded formulae for calculating costs of components and hence allow better and more rapid prediction of product cost.

5. Future Developments

5.1. Auto-Scheming

The KEE software is well suited to solving planning and scheduling problems and we intend to experiment with Schemebuilder by allowing it to automatically create schemes, 'domino fashion', but with designer intervention to apply filters and limits as each successive component is added to a set of competing schemes.

5.2. Layout Selection

It is intended to transfer the structure and methods developed in Creativity/Principles to help create and index a set of Layout rules that might be used both to guide the Layout process and to rank the competing layouts. The criteria used would be gathered under headings such as:

- Ease of Manufacture
- Cost of Manufacture
- Ease of Servicing
- Reliability

It is hoped that much of the information in such an index might be drawn from published work.[12,13]

6. Conclusion

This paper has described the underlying philosophy, approach to development, current status and future direction of the Schemebuilder and Layout projects of Lancaster University Engineering Design Centre. We believe that these tools show great promise and will make a significant contribution to the development of a truly *mechatronic* approach to product design, from the conceptual stage onwards.

7. References

1 French, M J, Bradley, D A and Dawson, D, "Proposal for the Establishment of an Engineering Design Centre - Case in Support", Lancaster University (1990)

2 Buur, J, "Positioning Mechatronics Design between Mechanics, Electronics and Software", Proc. Int. Conf. on Advanced Mechatronics, Tokyo, pp189-94 (1989)

3 Bradley, D A and Dawson D, "Information based strategies in the design of mechatronic systems", Design Studies, v12, n1, pp12-8 (Jan. 1991)

4 Buur, J and Andreasen, M M, "Design models in mechatronic product development", Design Studies, v10, n3, pp155-62 (July 1989)

5 Filman, R E, "Reasoning with Worlds and Truth Maintenance in a Knowledge-Based Programming Environment", Communications of the ACM v31, n4, pp382-401 (April 1988)

6 Paynter, H M, "Analysis and Design of Engineering Systems", MIT Press, Cambridge, MA, USA (1961)

7 Ward, A C & Seering, W P, "The performance of a mechanical design compiler", Proc. Int. Conf. on Engineering Design, London, pub. ASME, pp1273-85 (Aug. 1989)

8 Newell, R G and Parden, G, "Parametric design in the MEDUSA system", Proc. CAPE '83, 1st Int. Conf. Amsterdam, pp667-77 (April 1983)

9 Festo Ltd, Teddington, Middlesex, Pneumatics catalogue and disc.

10 Cramer, H, "Mathematical Methods of Statistics", Princeton University Press (1946)

11 French, M J, "Engineering Design: the Conceptual Stage", Heinemann, London (1971)

12 Andreasen, M M, Kahler, S and Lund, T, "Design for Assembly", IFS Publications (1983)

13 Boothroyd, G and Dewhurst, P "Product Design for Assembly", Boothroyd Dewhurst Inc, USA (1983)

A knowledge based approach to feature recognition in machining processes

B S Dalay, BEng, **R M Parkin**, BSc, PhD, CEng
Leicester Polytechnic, UK

SYNOPSIS The use of computers in modern manufacturing industry is widespread. Full exploitation of their potential benefit is often not achieved due to deficiencies in communication between different types of machine and file formats.

The design stage of the machine tool is critical in that it defines the operating regimes and data interchange processes. An enlightened Mechatronic approach at this stage is advantageous in providing a more elegant solution
and avoiding the 'Islands of Automation' syndrome.

The selection of an appropriate technique for the communication of product design description is a key factor in the formulation of the solution. The plotter communications language, HP-GL, has been universally adopted and offers many features which may be beneficially exploited.

The paper describes an HP-GL based communication system used on a multi-axis profiling tool grinder. In common with more conventional techniques of achieving a CADCAM link, the language definition fails to specify standards for its formal use. This is demonstrated via examples of the output from a range of CAD systems.

The development of knowledge based data interpolation, optimised off-line, is expected to provide a basis for a universal CAD interface. The language parsing components have provided error free operation in a generic plotter emulation system to asses various data parameters.

1 INTRODUCTION

The timber mill is an important facet of the woodworking industry. Their wide variety products is supported by a diverse range of machinery. Moulding and planing operations, in which the timber is fed through a machine housing multiple high speed rotary cutters, are fundamental to many production cycles. As a consequence the machines have been the subject of numerous refinements directed towards improving the output of quality machined timber.

The focus for much of the attention has been the tooling. This has traditionally taken the form of a square block which acts as a carrier for the cutting blades. The drift towards cylindrical block designs (see Fig 1), followed by the increases in tool spindle speeds to the now common 15000 rev/min(1), enabled the introduction of jointing. This technique involves an in-situ fine trimming of the cutting edges using an abrasive stone resulting in improved machining characteristics (2) and tool longevity(3). Although this provided some improvements, downtime through tool changes remained a problem which resulted in machine utilisation levels of less than 30%(1).

2 A MODULAR APPROACH TO CIM

A program of improvements to the setting procedures led to the development of an off-line measurement system, CAMMSET (Computer Aided Moulding Machine Setting System). This removed the trial and error from the setting procedure. An obvious advancement towards full CNC setting was however resisted by a market fearing failures found in other industries adopting CIM strategies with greater vigour(4).

Instead, the concept of CIM(5) was gradually introduced through the development of CASS (Computer Aided Setting System) (see Fig 2). Designed so that it may be retrofitted, to existing moulding machines, the system serves to present the operator with tool setting data for each of the many adjustment points. The communication with CAMMSET was achieved through either a serial data link or a smart card interface. The same interface also appears in, the AMMS (Automatic Moulding Machine Setting System), the optional extension module which further automates the process of machine setting by providing motorised adjustments.

These improvements highlighted deficiencies in the profile tool manual grinders. Again a phased introduction of a solution was adopted first with the ATG (Automatic Template Grinder), to simplify the production of templates for the existing base of profile grinders, followed by APG (Automatic Profile Grinder), a profile grinding machine modular extension.

3 THE PROBLEM OF COMMUNICATIONS

A criticism expressed against many early attempts to implement computer aided manufacturing was that they created 'islands of automation' (6). Communications was identified early in the development of ATG and APG as being critical in avoiding this problem and key to the successful implementation of CIM. A cost conscious design lead to a common control system architecture which featured a minimum user interface. This restricted its use for data entry describing the number of blades, the size of the cutterblock, the type of blade material and the angle with which they are held in the block. Communication of profile shape under this scheme would originate from an external source. Enhanced product marketability through independence of data source was the primary criteria used in the selection of the most appropriate technique.

The ideal scenario where the customer communicates his timber product demands electronically is, in practice, almost impossible to exploit here due to the variety of hardware and software systems. A number of standards have been formulated in an attempt to steer a path towards this goal of commonality in data exchange and interface. MAP (Manufacturing Automation Protocol) offered a solution covering both hardware and software interfaces. Its comprehensive nature was its major disadvantage in this application. STEP (Standard for the Exchange of Product model data) a system specifically conceived for product data interchange also suffers because it remains the subject of development(7).

IGES (Initial Graphics Interchange Specification) attempted to provide a standard for communicating CAD drawings(8). Although demonstrated as providing a transfer medium, its use is hampered by gaps in its definition(9) and poor levels of performance(10). DXF (Data Exchange Format), a proprietary system, is similar in purpose and levels of support.

HP-GL (Hewlett Packard Graphics Language) has established itself as a what many believe as the de facto standard for communicating with graphic pen plotters. Since this feature of producing an output is common across the whole spectrum of CAD hardware and software systems, it was found to offer the most desirable basis for profile data interchange.

4 HP-GL CHARACTERISTICS

HP-GL consists of over fifty core instructions. The command sequences largely consist of a two letter mnemonic followed by a list of numerical parameters. Appearing as conventional ASCII, each instruction is terminated by a separator such as a space, comma or semicolon. In addition to commands specifically concerned with the generation of the plot image, the language definition encompasses a comprehensive series of instructions dedicated towards the implementing efficient data transfer protocols.

Its use as a communications medium is complicated by a series of limitations:-

(a) Solid modelling - HP-GL provides no internal support for solid or three dimensional information representation. The data provides controls over pen movements according to a two axis cartesian coordinate system.

(b) Resolution - Instructions concerning pen movements and so drawing representation, use an addressing scheme have a resolution of 25 m. Subject to the limitations of conventional plotting mechanisms and media, such resolution is perfectly acceptable. However with both the ATG and APG designed with accuracy a primary design objective, the use of the language relies heavily on the ability to control the rational function relating drawing units to physical pen movements through scaling. This also acts to restrict the source of data to that which provides a scaled rather than a best fit plot output.

(c) Base unit definition - Early language definitions created some ambiguity from the references given to imperial unit equivalents. This has lead to a degree of uncertainty surrounding the interpretation software designers had adopted. Again limitations of conventional plotters help to disguise the problem but the more stringent interchange requirements emphasise even these subtle points.

(d) Consistency - HP-GL provides a common language definition across hardware and software boundaries. The language flexibility aids its primary role as with pen plotters but complicates its use as a profile data interchange medium.

(e) Entity support - The language provides little support for geometric entities such as splines and polygons. Instead these must be translated using low level commands to direct pen movements to discrete locations. This process results in a loss of entity information and produces a discontinuous set of data points. Again the limitations of the plotting mechanism assists in disguising the use of crude linear interpolation methods. High accuracy ATG and APG grinding operations in contrast would be unable to use the same techniques since they would be precisely resolved and later transferred on to the timber surface with detrimental effects on aesthetic appearances.

Some of these limitations are substantially offset by the advantages:

(a) Established uniform language definition which is common to the whole spectrum of computing hardware platforms and operating systems.

(b) Its universal support offers the user to benefit from the changing products in the

CAD market and the machine manufacturer from reduced machine development costs.

(c) Its basis of plotter hardware technology provides some assurance of continued future support by software developers.

(d) In addition to improving the marketability of the machines through the loose tie with the source of data, the technique is beneficial to the user in that use existing equipment simplifies the training procedures.

Both the ATG and APG would share a common initialisation procedure with the operator keying in data describing the machining operation. This would be followed by the generation of HP-GL data directly from the CAD environment using a simple scaled drawing of the timber shape. The communication would be initiated by the selection of the plot command which would transfer the data to the remote toolroom, using either an asynchronous serial or smart card control system interface.

5 HP-GL TRANSFER IMPLEMENTATION

Problems of consistency and the lack of comprehensive support of geometric entities are more difficult to solve when the machine requirements place constraints on mould shape preservation and accuracy. Lagrange polynomial techniques are simple to apply but results infringe both constraints. Piecewise curve generating mathematical techniques, such as those using splines or Bézier curves, are more complex to apply and produce similar results except that the control over curvature gives a more aesthetically pleasing result. Moulds that have mating faces and joints have a greater requirements for the accurate preservation of proportions and so would be more difficult to satisfy using these methods. The lack of consistency between the HP-GL generating mechanisms would hamper attempts to resolve these difficulties through fine tuning of the interpolation algorithms to arrive at continuous path contouring data suitable for the control system during the final machining phases.

6 A KNOWLEDGE BASED SOLUTION

A more flexible approach to a generic plotter emulation software interpolation engine, offering a universal CAD interface, is provided by the application of a knowledge based expert systems approach. Its purpose would be twofold:-

(a) Provide a plotter emulation interface which is independent of the CAD mechanism used for generating profile shape descriptions.

(b) To convert the command sequences and make decisions concerning profile shape between sampled data points extracted from the HP-GL codes.

Unlike applications of knowledge based expert systems in Medicine(11), Law(12) and the many disciplines of Engineering(13),(14),(15), the contents of the knowledge base, upon which decisions concerning profile shape would be based, cannot be easily identified. A machine learning system provides an alternative strategy which has proven its usefulness in solving numerical and logical problems(16). The primary objective of the learning procedure would be to identify and quantify heuristic operating parameters which gave an optimal performance when subjected to the complete spectrum of HP-GL data sources.

A major constraint on the design of the learning machine came from the use of an embedded control system architecture for both the ATG and APG machines. Modifications to the knowledge base as a consequence are limited once the operating system code has been established. This necessitated prior off-line construction and optimisation of the knowledge base together with its parameters.

7 OFF-LINE OPTIMISATION

The limitations of a fixed knowledge base in the target ATG and APG control systems acted to prevent its use as the basis for the learning machine. Instead a design was formulated in which use was made of a single personal computer platform, for which a variety of CAD software systems were readily available (see Fig 3). The main feature is an optimiser which compares the results and makes adjustments to the knowledge base. The iterative nature of the learning procedure dictated the design of the various software modules. This along with the eventual need to apply the results to the target system was reflected in first phase experiments to asses the suitability of various data measures (see Fig 4) to the software interpolation inference module.

8 PARAMETER IDENTIFICATION

A common source of input to both systems was a program designed to generate timber profile shapes upon demand. The data seed was a series of pseudo random numbers generated by a multiplicative congruential random number generator. To compensate for the lack of human input, data realism was controlled by the application of common rules limiting the shape of timber machined using a planer moulder. These included checks for acute angle features, excessive depths of profile and arc features too small to be reproduced using current profile grinding wheel technologies.

The random profile shape data structure was then recorded in a file for later comparison with the results of the knowledge based interpolator. The primary data path was through an integrated generic CAD command interpreter. Its role was to provide a sequence of keystroke commands that would enable the profile to be entered into a specified drawing editor. This was achieved in an unattended mode through an interrupt driven program to simulate keypresses. The final sequence of commands would save the drawing in the native format prior to generating a HP-GL data file.

The destination for the HP-GL code was a software language parser and plotter simulator. Its prominence in the final target system implementation again dictated the construction of the software which featured separated input output routines. A comprehensive plotter model was implemented to enable full emulation of responses critical in achieving a successful interface with more stringent CAD HP-GL data sources. These often make use of the extensive language support for communications protocols and error checking systems. In addition to providing output for a mimic display, a file containing a data structure of all the pen movement vectors received was generated.

The analysis module that followed in the suite of programs, first converted this data from the plotter base unit (1 unit = 25 m) to the metric units upon which the control system was based. The data was then scanned to determine the range of abscissa and ordinate values. Correction for the location of the coordinate system origin was then performed before the integrity was confirmed by checking for overlapping vectors and gaps in the between successive vectors.

The plotting algorithms in many CAD systems were found to generate HP-GL data following a sequence of drawing revision. More refined systems offered varying levels of pen movement optimisation directed towards minimised plot cycle times. In either case, the vectors were found to be of need of rearrangement to follow a consistent sequence before being further processed.

9 PRELIMINARY TESTS

Initial tests have revealed a diverse range of profile complexities. At one extreme simple profiles were found to be made up of a series of straight line segments (see Fig 5). Processing the results through a selection of three differing CAD systems produced similar vector descriptions (see Table 1). Small deviations of were observed in the points used to define the five vectors in the example. The distribution of errors were at a maximum towards the end of the profile width in all but the case of Designcad 2D result. Instead this system gave errors consistently within the bounds of a 25 m plotter unit.

All of the systems were found to issue integer rather than floating point pen movement parameters which are within the scope of the language definition. This helps explain a common error component. Other sources of error may have arisen from the internal precision of the various drawing editors and the relationship with command line drawing inputs. The random profile itself was generated to a precision of seven digits. This contrasts many of the inputs being restricted to six significant digits. The results from the Cadkey test indicate an accumulation of errors which suggests the use of a relative mode of pen movement in the code generating algorithms.

A more sophisticated profile shape made of three arc sections shows a contrasting vector sequence (see Fig 6). Although the language supports high level commands to instruct the plotter to produce arc and circle features, none of the systems took advantage of its compact format. Instead the action of the command was emulated using lower level commands to move the pen to discrete locations. These described a number of tightly spaced chords which under the finite mechanical resolution of the pen plotter would normally produce a good shape approximation.

As an aside, a small experiment in which circles of varying diameters were plotted from two differing drawing editors was performed to further investigate the nature of the vector chords. The results (see Fig 7) showed a trend towards increased numbers of chords with increased circle diameter. Clipping as a result of attempts to move the pen outside the plot window explain some of the differences found. The differing rates of change in the number of chords is more difficult to attribute to factors other than fundamental differences in internal HP-GL code generating algorithms. Such differences are justification for the use of knowledge based techniques.

As opposed to observing the absolute measures in such schemes greater success is expected from an analysis of trends and enlightened use of historical results. An example of such a parameter measure may be found in the magnitude of the vectors describing the chord features on the profile outline (see Fig 8). The results obtained through the analysis software module show a clear pattern of near constant magnitude in profile regions described by arcs. The level of the vector magnitude also indicates the relative sizes of the radii. Such changes are expected to be of benefit in identifying the junction between differing entity types.

10 CONCLUSION

A conservative timber mill machinery market has dictated the development of systems to enhance the production of quality machined timber. The planer moulder machine due to its importance in many manufacturing cycles has been the centre of development attention. A modular approach has resulted in several retrofit systems which have improved machine utilisation.

Further significant gains may be expected from the automation of the tool preparation techniques. A two phase development strategy has identified communications as the key to achieving a successful cost conscious design which adheres to CIM philosophies. Consideration at the design stage has enabled the features of HP-GL as a profile shape transport medium to be analyzed.

Its primary use as a plotter language has governed the specification for the machine interface with the CAD environments. The use of knowledge based techniques is expected to enable the differences in code generating algorithms to be accommodated, thus providing a universal means for the control system to accept data describing the machining processes. The use of embedded control system architectures has necessitated the use of an off line iterative learning machine approach to establish the heuristic operating parameters.

Test results have identified clear trends in results from a number of drawing packages. Measures such as vector magnitude have been clearly shown to be good indicators of changes in profile shape. Further tests are required to quantify the reliability of this measure across a range of CAD data sources.

REFERENCES

(1) Parkin R.M. The application of CIM to the planing and moulding machines in the wood working industry. COMODEM 88 The first U.K. seminar on condition monitoring and diagnostic engineering management. City of Birmingham Polytechnic, Birmingham, 19th-21st September 1988.

(2) Maycock K.M., Parkin R.M., Buttery T.C. A sensor for in-process surface form measurement of planed timber. Eurosensors, The third conference on sensors and their application, Cavendish Laboratory, Cambridge 22nd-24th September 1987.

(3) Jackson M.R. Some of machine characteristics on the surface quality of planed and spindle moulded wooden products. CNAA PhD thesis, June 1986, Leicester Polytechnic.

(4) Shaiken H. The human impact of automation, keynote speech to the 1985 conference on decision and control. Institute of Electrical and Electronic Engineers Control Systems Magazine, December 1986, pp 3-9.

(5) Gunn T.G. CAD/CAM/CIM: now and in the future. Instrumentation and Control Systems, Vol 58, No 4, April 1985, pp 59-64.

(6) Astrop A. Integration the crucial challenge. Machinery and Production Engineering, Vol 142, Pt 3652, pp 38-40, 1984.

(7) Schlechtendahl E.G., Gengenbach U. CAD data transfer: The goals of project 322 CAD interfaces (CAD*I). Esprit'87 Achievements and impacts, Proceedings of the 4th annual Esprit conference, Brussels, 28th-29th September 1987.

(8) Smith B., Rinaudot G.R., Reed K.A., Wright T. Initial graphics exchange specification (IGES) version 4.0. National Bureau Of Standards, U.S. Department of Commerce June 1988, Document No. NBSIR 88-3813.

(9) Dobson R.S. Effective CADCAM data exchange. Effective CADCAM 1985 Fitzwilliam College, Cambridge, 8-10 July 1985, IMechE Conference Publication.

(10) Gane R.D. The link between CAD and CAM Planning for Automated Manufacture Proceedings of the IMechE, Coventry 24-25 September 1986.

(11) Shortcliffe E.H. Computer-Based Medical consultation: MYCIN, Published by American Elsevier 1976.

(12) Botten N., Raz T. Development of the knowledge base of an expert system for civil rights investigation. Expert Systems Vol 7, No 4, Nov 1990, pp 224-235.

(13) Miysato G.H., Dong W., Levitt R.E., Boissonnade A.C. Implementation of a knowledge based seismic risk evaluation system on microcomputers. International Journal For Artificial Intelligence in Engineering, Vol 1, No 1, July 1986, pp 29-35.

(14) Wright M.L., Green M.W., Fiegl G., Cross P.F. An expert system for real-time control. IEEE Software, Vol 3, No 2, Mar 1986, pp 16-24.

(15) Turban E. Expert systems-based robot technology. Expert Systems, Vol 7, No 2, May 1990, pp 102-110.

(16) Wang Wu and Chen Jianhua, Learning by discovering problem solving heuristics through experience. IEEE Transaction On Knowledge And Data Engineering, Vol 3, No 4, December 1991, pp 415-420.

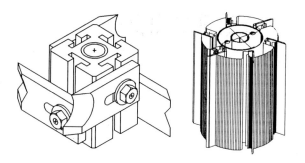

Fig 1 Square cutterblock and its replacement cylindrical type which offers improved performance

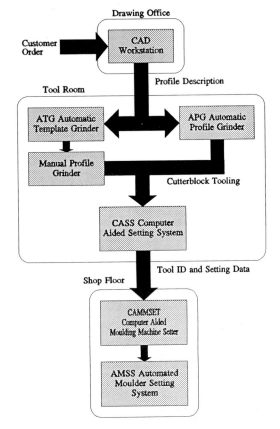

Fig 2 An illustration of the modular approach to CIM for the timber mills

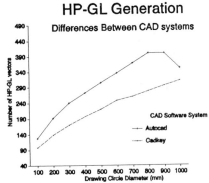

Fig 3 Variation of HP–GL vector numbers with circle diameter

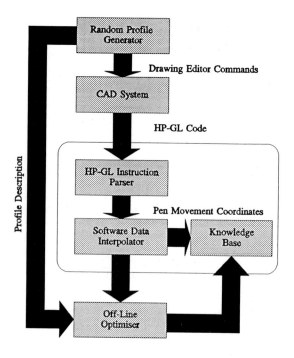

Fig 4 Off-line iterative optimization of the knowledge based interpolator

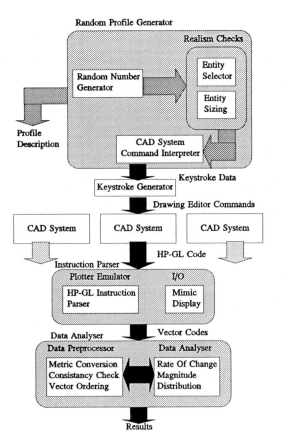

Fig 5 Identification of knowledge base heuristic parameters

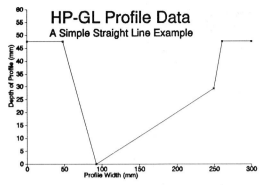

Fig 6 A simple five vector outline produced by the random profile generator

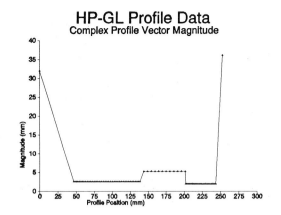

Fig 8 HP–GL vector data magnitude as an indicator of profile shape

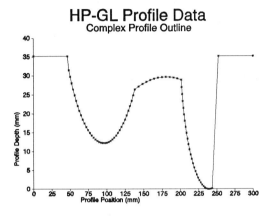

Fig 7 A complex profile outline consisting of three arc sections

Profile	Input Drawing					
Width (mm)	0	48.18	92.59	249.2	260.73	300
Depth (mm)	47.59	47.59	0	29.21	47.59	47.59
	HP-GL Profile Data Errors					
	Autocad Ver 10.0					
Width (mm)	0	-0.005	-0.015	-0.025	-0.03	-0.025
Depth (mm)	-0.015	-0.015	0	-0.01	-0.015	-0.015
	Cadkey Ver 1.4E					
Width (mm)	0	+0.02	+0.035	+0.075	+0.07	+0.1
Depth (mm)	-0.025	-0.025	0	+0.025	-0.025	-0.025
	Designcad 2D ver 4.2					
Width (mm)	0	-0.005	-0.01	0	-0.005	0
Depth (mm)	-0.015	-0.015	0	-0.01	-0.015	-0.015

Table 1 HP-GL Profile Vector Error Comparison

Design of mechatronic systems using bond graphs

P J Gawthrop, MA, DPhil, MIEE, MInstMC, CEng, EurIng
Glasgow University, UK.

SYNOPSIS. Bond graphs are used as the basis for a computer-based set of modelling tools particularly appropriate for the design of mechatronic systems. Their use is illustrated by the design of a control system for a robot.

1 Introduction

Good design needs good models, and thus mechatronic design requires a good modelling technique applicable across a range of physical domains; bond graphs [1][2][3][4][5][6] provide the basis for such a representation.

Modelling of dynamic systems in general, and robots in particular, is difficult. The modelling process can, however, be helped by interaction with appropriate computer-based modelling tools. In the field of robotics, there have been a number of such tools developed [7] [8] [9] [10] [11] [12] [13] [14] [15].

Here, I use the word 'metamodelling' to describe the generic techniques lying behind the modelling of specific systems. A current research programme at Glasgow, is concerned with:

- elucidating the foundations of metamodelling [16][17];
- generating computer-based metamodelling support tools [18] [19];
- applying the methods to a range of dynamic systems including:
  - mechatronic systems including industrial robots [20]
  - chemical processes [21]
  - pharmacokinetics
  - macroeconomics

The research has focussed on the Bond Graph technique [1][2][3][4][5][6] for reasons given elsewhere [16][17].

Modelling is viewed as an iterative and interactive process: the software tools help the user comprehend a complex system; the user points the software in appropriate directions. There a wide range of system *representations* which the user may wish to view and the software tools provide a means of *transforming* between such representations [19].

To illustrate the power and use of the methods, this paper describes a case study involving a significant mechatronic system: a large industrial robot. For simplicity in illustrating our approach, non-ideal properties such as compliance and friction are beyond the scope of the paper; but the non-linear rigid-body dynamics are included. However, the model is written in such a way that the non-ideal effects may be easily included by adding appropriate components to the bond graph. The dynamics are restricted to 2D motion in the horizontal plane.

The following representations of the model are *automatically* derived as the modelling process evolves.

- Differential/algebraic equations (section 3).
- Constrained state equations (section 4).
- Linearised constrained-state equations (section 5).
- Simulation of the non-linear system (section 6.2).
- Frequency response of the linearised system (section 6.3).
- Root locus of the linearised system under PD control (section 6.4).
- Simulation of the linearised constrained-state equations (section 6.5).

In addition, the following control laws (section 6) are evaluated:

- independent PD control of each loop,
- a 2×2 decoupling PD control,

The performance of the controllers is examined when:

- the load mass is changed
- the arm configuration is changed.

All models were automatically generated using MTT [19], and the numerical analysis and simulation, together with the graphics, were performed in Matlab.

2 Bond graph model

The bond graph describing the robot appears in Figure 1.

The modelling technique follows that in reference [20], but the following points should be noted.

1. For those readers unfamiliar with the bond graph technique, it suffices to realise that Figure 1 provides a precise and unambiguous description of the robot dynamics.

2. The two elements labeled **jmo** and **jm2** represent the two motor inertias. These have been assigned integral causality; as it turns out, this implies that all other inertias have derivative causality.

3. Individual components of the robot have been modelled, thus masses and inertials can be directly included in the model.

See end of paper for Figure 1

Figure 1: Industrial robot: bond graph

4. Although the model contains no compliance or friction, these components may be directly added to the appropriate junctions. This is one reason for its apparent complexity.

5. The motor angles and the joint angles have separate states associated with them. This has the disadvantage of adding two extra states but has the following advantages:

   (a) The motor and joint angles are physically distinct quantities and so should have separate components associated with them.

   (b) When gearbox compliance is included, the angles are no longer dependent.

   (c) *relative* angles can be used at the motors; *absolute* angles can be used at the joints.

6. A point mass at the tip of the extension tube has been included.

7. Each link is represented by *physical* parameters: a mass m_i, an inertia j_i, joint – mass centre distance a_i, mass centre – joint distance b_i and total distance between joints l_i. These are initially in *symbolic* form. They are then specialised to the actual numerical values as appropriate.

3 Differential-algebraic equations

As discussed elsewhere [20], the equations describing a manipulator may be readily derived in differential-algebraic form. The algebraic equations arise from the constraints between the rotational and translational motion of the links imposed by the transformers in Figure 1.

The some of the automatically generated equations are given below; the full set is too long to be included here. θ_0 and θ_2 are the motor angles; α_0 and α_2 are the link angles; h_{m0} and h_{m2} are the motor angular momenta; h_i is the ith link angular momentum; $m_i \dot{x}_i$ is the ith momentum of the centre of mass of the ith link; and λ_0 and λ_2 are the flux linkages of the two motors.

$$x = \begin{pmatrix} \theta_0 \\ \theta_2 \\ \alpha_0 \\ \alpha_2 \\ h_{m0} \\ h_{m2} \end{pmatrix} ; \; z = \begin{pmatrix} h_0 \\ h_1 \\ m_1 \dot{x}_1 \\ m_1 \dot{y}_1 \\ h_2 \\ m_2 \dot{x}_2 \\ m_2 \dot{y}_2 \\ m_3 \dot{x}_3 \\ m_3 \dot{y}_3 \\ \lambda_0 \\ \lambda_2 \end{pmatrix} ; \; y = \begin{pmatrix} \theta_0 \\ \theta_0 \\ \dot{\theta}_0 \\ \dot{\theta}_2 \end{pmatrix} ; \; u = \begin{pmatrix} i_0 \\ i_2 \end{pmatrix} \quad (1$$

$$\dot{x}_1 = \frac{x_5}{j_{m0}} \tag{2}$$

$$\dot{x}_2 = \frac{x_6}{j_{m2}} \tag{3}$$

$$\dot{x}_3 = \frac{x_5}{(j_{m0} n_0)} \tag{4}$$

$$\dot{x}_4 = \frac{(j_{m0} n_0 x_6 + j_{m2} n_2 x_5)}{(j_{m0} j_{m2} n_0 n_2)} \tag{5}$$

$$z_1 = \frac{(j_0 x_5)}{(j_{m0} n_0)} \tag{6}$$

$$z_2 = \frac{(j_1 x_5)}{(j_{m0} n_0)} \tag{7}$$

This form of equation representation is fairly standard in numerical solution of multi-body problems; and, in some cases, solvers are available to the numerical simulation of such sets of equations [22].

From the way that these equations arise from the bond graph, the variables are clearly divided into 6 states ($x_1 - x_6$) and 11 other variables ($z_1 - z_{11}$) that would be states except that they are constrained in terms of the 6 states; these will be called *non-states*.

Both sets of variables have clear physical significance in terms of momenta. The choice of this division is up to the modeller, and is directly related to the bond graph causality shown in Figure 1: I elements with integral causality become states, those with derivative causality become non-states. Because the state equations are *linear* in the non-state derivatives $\dot{z}_i$, the equations can be rewritten in a more useful form; this is accomplished in Section 4.

4 Constrained state-space equations

In reference [20], it is shown that the DAEs of section 3 can be rewritten as

$$\dot{\chi} = AX + Bu \tag{8}$$

where

$$\dot{\chi}(X) = E\dot{X} \tag{9}$$

In this case, the corresponding equations are

$$\dot{\chi}_1 = \frac{x_5}{j_{m0}} \tag{10}$$

$$\dot{\chi}_2 = \frac{x_6}{j_{m2}} \tag{11}$$

$$\dot{\chi}_3 = \frac{x_5}{(j_{m0} n_0)} \tag{12}$$

$$\dot{\chi}_4 = \frac{(j_{m0} n_0 x_6 + j_{m2} n_2 x_5)}{(j_{m0} j_{m2} n_0 n_2)} \tag{13}$$

$$\dot{\chi}_5 = k m_0 u_1 \tag{14}$$

$$\dot{\chi}_6 = k m_2 u_2 \tag{15}$$

$$y_1 = x_1 \tag{16}$$

$$y_2 = x_2 \tag{17}$$

$$y_3 = \frac{x_5}{j_{m0}} \tag{18}$$

$$y_4 = \frac{x_6}{j_{m2}} \tag{19}$$

The E matrix is too complicated to include here.

5 Linearised constrained-state equation

The equations in section 4 can be linearised (about a given state X_0) to give a set of *linear constrained-state equations* in the form:

$$E(X_0)\dot{X} = AX + Bu \qquad (20)$$

By symmetry, θ_0 can be taken as zero without loss of generality. If, in addition, the nominal velocities are taken to be zero then:

$$X_0 = \begin{pmatrix} 0 \\ \theta_2 \\ 0 \\ 0 \end{pmatrix} \qquad (21)$$

The advantage of a symbolic representation with respect to a numerical representations that the effect of individual components can be directly revealed; a disadvantage is that, as in the previous section, expressions may be rather long. A useful compromise is to leave the quantities of interest as symbols but set the other symbols to numerical values.

As an example of this, is is of interest to examine the effect of the gearbox ratios on the non-linear terms of the matrix E. All symbols except θ_2, n_0 and n_2 are set to their numerical values (suitably rounded). The linearised equations then become:

$$A = \begin{pmatrix} 0 & 0 & 0 & 0 & 20 & 0 \\ 0 & 0 & 0 & 0 & 0 & 21.7391 \\ 0 & 0 & 0 & 0 & \frac{20}{n_0} & 0 \\ 0 & 0 & 0 & 0 & \frac{20}{n_0} & \frac{21.7392}{n_2} \\ 0 & 0 & 0 & 0 & 0 & 0 \\ 0 & 0 & 0 & 0 & 0 & 0 \end{pmatrix} \qquad (22)$$

$$B = \begin{pmatrix} 0 & 0 \\ 0 & 0 \\ 0 & 0 \\ 0 & 0 \\ 0.999997 & 0 \\ 0 & 1.9 \end{pmatrix} \qquad (23)$$

$$C = \begin{pmatrix} 1 & 0 & 0 & 0 & 0 & 0 \\ 0 & 1 & 0 & 0 & 0 & 0 \\ 0 & 0 & 0 & 0 & 20 & 0 \\ 0 & 0 & 0 & 0 & 0 & 21.7391 \end{pmatrix} \qquad (24)$$

$$D = \begin{pmatrix} 0 & 0 \\ 0 & 0 \\ 0 & 0 \\ 0 & 0 \end{pmatrix} \qquad (25)$$

$$\begin{aligned}
e_{11} &= 1 \\
e_{22} &= 1 \\
e_{33} &= 1 \\
e_{44} &= 1 \\
e_{55} &= \frac{(18750.0(cos(\theta_2) + 0.53333310^{-6} n_0^2 + 2.27467))}{n_0^2} \\
e_{56} &= \frac{(10190.2(cos(\theta_2) + 0.317333))}{(n_0 n_2)} \\
e_{65} &= \frac{(9375(cos(\theta_2) + 0.317333))}{(n_0 n_2)} \\
e_{66} &= \frac{(n_2^2 + 3233.7)}{n_2^2}
\end{aligned} \qquad (26)$$

Only non-zero elements of the E matrix have been given. As the gearbox ratios are large in this case (about 100) the variation of E with θ_2 is not large.

6 Control

To illustrate how these models can be used for the purposes of control design. In particular four stages of design and analysis are used for illustration.

- controller-specific models are automatically generated (Section 6.1)
- non-linear simulation code is automatically generated (Section 6.2)
- Matlab code is automatically generated for analysing the design using both frequency response and root-locus methods. (Sections 6.3 and 6.4)
- linearised simulation code is automatically generated (Section 6.5)

6.1 Control design

The paper investigates two controllers:

- independent PD control of each loop,
- a 2×2 decoupling PD control,

The linearised model of the (rigid) robot is essentially a multivariable double integrator of the form:

$$G(s) = \frac{M_2}{s^2} \qquad (27)$$

where M_2 (the second Markov parameter [23]) is a 2×2 matrix:

$$M_2 = \begin{pmatrix} m_{11} & m_{12} \\ m_{21} & m_{22} \end{pmatrix} \qquad (28)$$

The following simple design approach is used for the decoupled controller

1. Using MTT, generate the linearised constrained-state model, equation 20.

2. Compute

$$A_e = E^{-1}A; \quad B_e = E^{-1}B \qquad (29)$$

and hence

$$M_2 = C A_e B_e \qquad (30)$$

3. Precompensate the system with M_2^{-1} to give two decoupled unit integrators

$$G(s)M_2^{-1} = \frac{I_{2\times 2}}{s^2} \qquad (31)$$

4. Design the PD control to give two closed loop poles with time constant T. ($T = 0.2 sec$ is used in this paper).

The two independent PD controllers are designed by the same method except that M_2 is replaced by the diagonal matrix:

$$M_{2d} = \begin{pmatrix} m_{11} & 0 \\ 0 & m_{22} \end{pmatrix} \qquad (32)$$

In all cases the model was linearised about $\theta_1 = \theta_2 = 0$.

6.2 Simulation of the non-linear system

Non-linear systems may be linearised (about a given state) to give a *linear* set of differential equations. Such linearised models have the advantage of being easy to analyse, simulate and are amenable to linear control synthesis (see for example Section 6). On the other hand, they are only an approximation to the non-linear system and, as such, may give misleading results.

This section provides both linear and non-linear simulation of the manipulator using the decoupling PD controller of section 6.1.

Figure 2 is a simulation of the linearised model (linearised about $\theta_1 = \theta_2 = 0$) for a change in setpoint of $\theta_1 = \frac{\pi}{2}$ and $\theta_2 = \frac{\pi}{4}$ and superimposed on the setpoint.

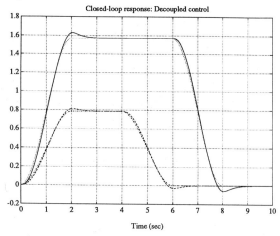

Fig 2 Linear simulation

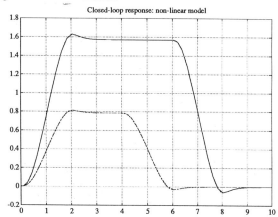

Fig 3 Nonlinear simulation: small movement

Figure 3 gives the corresponding non-linear simulation (superimposed on the linear result) Not surprisingly, the two simulations give identical results: a $\frac{\pi}{2}$ change in motor angle gives a very small change in joint angle. This result thus serves to verify the linearisation.

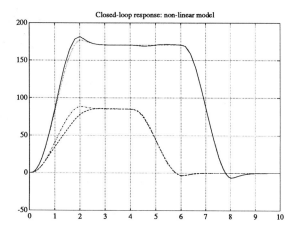

Fig 4 Nonlinear simulation: big movement

More significantly, Figure 4 shows the same non-linear simulation (superimposed on the linear result) except that the setpoint has been magnified by the gearbox ratio on joint 0. Thus the *joint* angle change is now $\frac{\pi}{2}$. There is now a discrepancy between the linear and non-linear result; but this is small enough to give confidence in using linear design and analysis before using non-linear simulation for a final check.

6.3 Frequency response of the linearised system

For the purposes of multivariable design (see, e.g. Maciejowski [24] [25]), it is convenient to look at the frequency response of the two input – two output system return ratio with a PD compensator in place. In other words, the compensator matrix K is appended to the system output where:

$$K = \begin{pmatrix} k_{p0} & 0 & k_{d0} & 0 \\ 0 & k_{p2} & 0 & k_{d2} \end{pmatrix} \qquad (33)$$

together with the precompensator. The configuration appears in Figure 5.

The corresponding Nyquist frequency responses appear in Figures 6 – 9. Figure 6 corresponds to the *decoupled* control of section 6; figure 7 corresponds to the *decoupled* control of

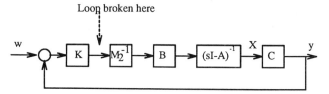

Fig 5 Control configuration

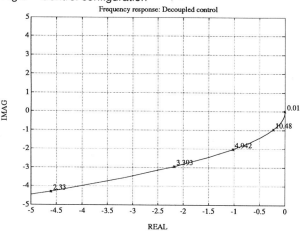

Fig 6 Frequency response: decoupled control

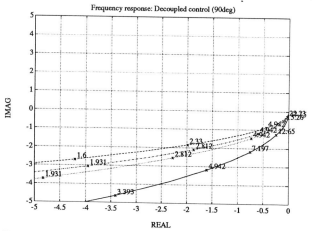

Fig 7 Frequency response: decoupled control (90 degrees)

section 6, but with a movement about a $\frac{\pi}{2}$ joint angle; figure 8 corresponds to the *independent* control of section 6; and figure 9 corresponds to the *decoupled* control of section 6 but with zero tip mass.

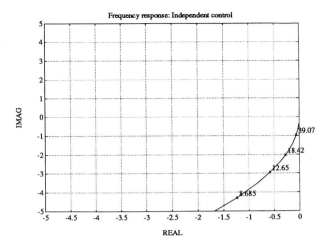

Fig 8 Frequency response: independent control

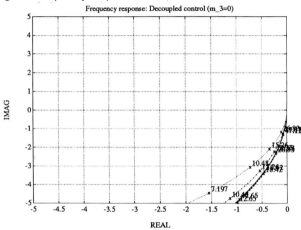

Fig 9 Frequency response: decoupled control (zero tip mass)

6.4 Root locus of the linearised system under PD control

In a similar fashion to section 6.3, the root locus of the compensated 2 input – two output system can be derived and are plotted in Figures 10 – 13. The roots marked '+' correspond to the nominal controller design, the other root marked '.' correspond to gains between 10^{-2} and 10^2 nominal.

Figure 10 corresponds to the *decoupled* control of section 6; the nominal poles are both at the designed value of -5. Figure 11 corresponds to the *decoupled* control of section 6, but with a movement about a $\frac{\pi}{2}$ joint angle; the two poles corresponding to the nominal gain are now complex corresponding to an overshooting response. Figure 12 corresponds to the *independent* control of section 6; the two poles corresponding to the nominal gain are now complex corresponding to an overshooting response. Figure 13 corresponds to the *decoupled* control of section 6 but with zero tip mass.

6.5 Simulation of the linearised constrained-state equations

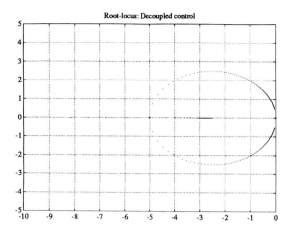

Fig 10 Root locus: decoupled control

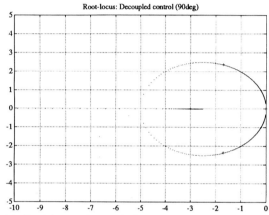

Fig 11 Root locus: decoupled control (90 degrees)

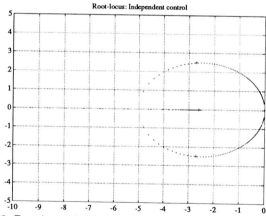

Fig 12 Root locus: independent control

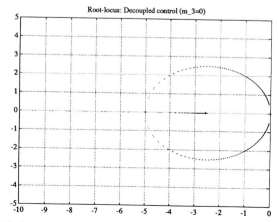

Fig 13 Root locus: decoupled control (zero tip mass)

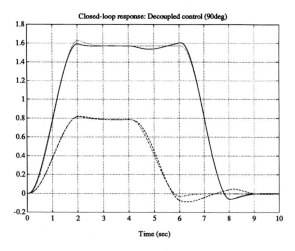

Fig 14 Linear simulation: decoupled control (90 degrees)

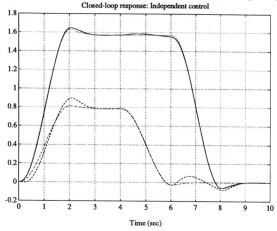

Fig 15 Linear simulation: independent control

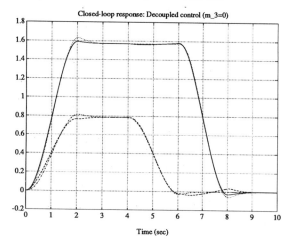

Fig 16 Linear simulation: decoupled control (zero tip mass)]

Linear simulations of the four situations analysed in sections 6.3 and 6.4 are shown in Figures 2 – 16.

Figure 2 corresponds to the *decoupled* control of section 6; figure 14 corresponds to the *decoupled* control of section 6, but with a movement about a $\frac{\pi}{2}$ joint angle; figure 15 corresponds to the *independent* control of section 6; and figure 16 corresponds to the *decoupled* control of section 6 but with zero tip mass.

These simulations fulfil the root-locus predictions. In each case, the result is superimposed on the linear simulation with decoupled controller of Figure 2.

7 Conclusion

The industrial robot has been successfully modelled using the generic bond-graph based metamodelling approach, and a preliminary control design study accomplished.

It appears that the large gearbox ratios considerably reduce the non-linearities due to rigid body motion. Thus the control problem reduces to the control of two (coupled) dc motors, the connection of these to the manipulator does not have a large effect. This means that control studies can be accomplished on the linearised model, and then checked against the full non-linear simulation.

The power of the metamodelling approach is illustrated by the relative ease by which the bond graph description could be automatically transformed into other representations. These transformations (including linearisation) are accomplished symbolically. Thus they can be checked by hand and can give insight into the equation structure; see, for example, the discussion in section 5. The equations are also algebraically simplified at this stage reducing later numerical computation.

Matlab was used the purposes of control system design, analysis and simulation, the necessary system descriptions (linear and nonlinear) where automatically generated. This gives access to the wide range of control design and analysis tools available in this environment.

Future work based on this model will include:

1. the modelling and analysis of gearbox compliance,
2. the modelling and analysis of link compliance,
3. the modelling and analysis of friction,
4. the validation of the model against the industrial robot,
5. the evaluation of advanced control strategies.

8 Acknowledgements

The metamodelling of robots is supported by SERC through the ACME directorate; the bond-graph tools (MTT) were developed though the SERC supported Engineering Design Research Centre at Glasgow.

References

[1] H. M. Paynter, *Analysis and design of engineering systems*, MIT Press, Cambridge, Mass., 1961.

[2] D. C. Karnopp and R. C. Rosenberg, *System Dynamics: A Unified Approach*, John Wiley, 1975.

[3] P. E. Wellstead, *Introduction to Physical System Modelling*, Academic Press, 1979.

[4] R. C. Rosenberg and D. C. Karnopp, *Introduction to Physical System Dynamics*, McGraw-Hill, 1983.

[5] J. Thoma, *Introduction to Bond Graphs and their Applications*, Pergamon Press, 1975.

[6] J. U. Thoma, *Simulation by bond graphs*, Springer-Verlag, Berlin, 1990.

[7] L. G. Herrera-Bendezu, E. Mu, and J. T. Cain, Symbolic Computation of Robot Manipulator Kinematics, in *IEEE Int. Conf. on Robotics and Automation*, pages 993–998, 1988.

[8] A. P. Tzes, S. Yurkovich, and F. D. Langer, A Symbolic Manipulation Package for Modeling of Rigid or Flexible Manipulators, in *IEEE Int. Conf. on Robotics and Automation*, pages 1526–1531, 1988.

[9] S. Cetinkunt and W. J. Book, Symbolic Modeling of Flexible Manipulators, in *IEEE Int. Conf. on Robotics and Automation*, pages 2074–2081, 1987.

[10] G. Cesareo, F. Nicolo, and S. Nicosia, DYMIR:A Code for Generating Dynamic Model of Robots, in *IEEE Int. Conf. on Robotics and Automation, Atlanta*, 1984.

[11] M. Vukobratovic and N. Kirkanski, A Method for Computer-Aided Construction of Analytical Models of Robotic Manipulators, in *IEEE Int. Conf. on Robotics and Automation, Atlanta*, 1984.

[12] L. Vecchio, S. Nicosia, and F. Nicolo, Automatic Generation of Dynamical Models of Manipulators, in *Proc. 10th ISIR, Milan*, 1980.

[13] J. J. Murray and C. P. Neuman, ARM: An Algebraic Robot Dynamic Modelling Program, in *IEEE Int. Conf. on Robotics and Automation*, pages 103–114, 1984.

[14] M. Vukobratovic, Computer-Aided Generation of Manipulator Kinematic Models in Symbolic Form, in *15th ISIR*, pages 1043–1049, 1985.

[15] J. F. Malm, Symbolic Matrix Manipulation with LISP Programming, in *Robots 8 — Conf. Proc., Robotics Int. of SME, Detroit*, pages 20.1–20.19, 1984.

[16] P. J. Gawthrop and L. Smith, Causal Augmentation of Bond Graphs, *Journal of the Franklin Institute*, (to appear).

[17] P. J. Gawthrop and L. Smith, Inverse Systems: Bond Graph and Descriptor Representations, *Journal of the Franklin Institute*, (submitted).

[18] P. J. Gawthrop and L. Smith, An environment for Specification, Design, Operation, Maintenance, and Revision of Manufacturing Control Systems, in *Proceedings of UKIT90*, pages 104–110, 1990.

[19] P. J. Gawthrop, N. A. Marrison, and L. Smith, MTT: A bond graph toolbox, in *Proceedings of the 5th IFAC/IMACS Symposium on Computer-aided Design of Control Systems:CADCS91, Swansea, Wales*, pages 274–279, 1991.

[20] P. J. Gawthrop, Bond Graphs: A representation for Mechatronic Systems, *Mechatronics*, 1(2):127–156, April 1991.

[21] S. A. MacKenzie, P. J. Gawthrop, R. W. Jones, and J. W. Ponton, Systematic Modelling of Chemical Processes, in *IFAC symposium on Advanced Control of Chemical Processes, ADCHEM'91. Toulouse, France.*, 1991.

[22] K. E. Brenan, S. L. Campbell, and L. R. Petzold, *Numerical solution of Initial Value Problems in Differential-algebraic equations*, North-Holland, New York, 1989.

[23] T. Kailath, *Linear Systems*, Prentice-Hall, 1980.

[24] J. M. Maciejowski, *Multivariable Feedback Design*, Addison-Wesley, 1989.

[25] M. P. Ford, J. M. Maciejowski, and J. M. Boyle, *Multivariable Frequency Domain Toolbox User's Guide*.

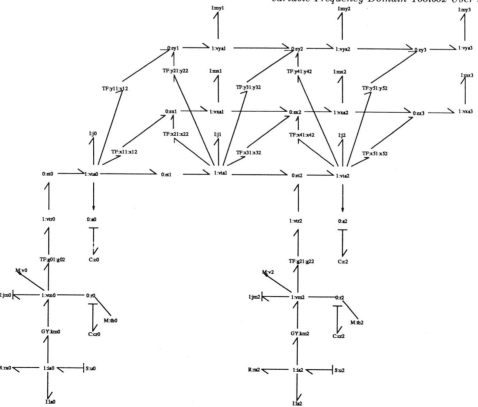

Fig 1   Industrial robot: bond graph

Design function deployment; a platform for cross - disciplinary product development

A JEBB, BSc, PhD, DIC, R C EDNEY, BSc, PhD, DIC, ACGI, S SIVALOGANATHAN, BSc, MSc, PhD,
N F O EVBUOMWAN, BEng, MSc, DIC
City University, London, UK.

SYNOPSIS *The design of a product always commences with the identification of the need for it, irrespective of whatever means the need is arrived at. In the present industrial climate, the success of a product often depends on the integration of electronics, control engineering, computing technologies and other design disciplines. This process of integration in engineering design, is what has come to be known as 'Mechatronics', and it involves the use of enabling technolgies such as sensors, actuators, software, design principles and communication. This integration requires design aids and tools in order to achieve an integrated product.*

In this paper, Design Function Deployment (DFD), a formal system to be implemented as a computer based system and which constitutes a natural platform for such integrated and cross-disciplinary product development is described. DFD enables the systematic translation of explicit and implicit customer requirements into design requirements (functions). These design functions constitute the criteria for the proposal of candidate designs and eventual optimization and selection of most promising mechatronic design.

1.0 INTRODUCTION

Irrespective of whether based on market analysis or on the designer's ingenuity, the process of product design starts with the identification of the need for the product. The success of any present day product often depends on the integration of electronics and computing technologies, together with other design disciplines, into the development process. A properly conceived product exploits the development in technology at the design stage in both product and manufacturing process design.

Mechatronics is the synergetic combination of precision mechanical engineering, electronic control and system thinking in the design of products and processes [1]. Present day products are characterised by the optimal use of electronics and control engineering, and the requirements on the physical hardware are often taken for granted. For instance when consumers set out to buy washing machines, they often look out for the different programs that are available (electronics and control engineering parts) rather than the structure of the machine or the power of the motor that runs it. Such products which are sometimes called mechatronics products, are not just combinations of electronic and mechanical hardware, but they are products developed by the integration of electrical, electronics and mechanical engineering. This integration demands design aids and tools in these disciplines which have to be available to the design team to achieve an integrated product.

Traditionally, product development is performed in a sequential manner, whereas the modern and more progressive approach is to consider at the design stage all the factors such as cost, manufacturing processes, reliability, safety, statutory requirements, etc that affect the design of the product, in a parallel manner.

This paper describes Design Function Deployment (DFD) a system which incorporates the techniques of Quality Function Deployment (QFD). It is an overarching system with facilities to hook a number of engineering and scientific programs to support the development of a cross-disciplinary or integrated product in a concurrent manner.

2.0 MECHATRONICS

Mechatronics essentially involves the integration of electronics and computer science into mechanical products and process design. It's concept underlines the necessity for effective and meaningful interaction between different fields of engineering design. The performance of products and their manufacture depends largely on the capacity of industry to exploit developments in technology. In most innovative products and processes, the mechanical hardware is that which first seizes the imagination and attention of designers, but the best realization usually depends on the consideration of the necessary electronics, control engineering and computing aspects from the earliest stage of the design process [1].

A number of enabling technologies including sensors, actuators, software, design principles and communication are involved in the integration of the Mechatronics concept in product/process development. Mechatronics is not a new branch of engineering, but a newly developed concept that underlines the necessity for powerful interaction between diferent fields of engineering [2].

In order to be successful, a mechatronics approach needs to be established from the very early stages of the conceptual design process, where options can be kept open (as provided for within the Design Function

Deployment System) before the embodiment is determined. This allows the designer to look at many if not all possible options.

2.1 CHARACTERISTICS OF A MECHATRONICS SYSTEM

Mechatronics approach to engineering design results in simpler systems with fewer components than the traditional purely mechanical designs. This is due to the transfer of complex functions to electronics and a good design system should enable the designer to look at these transfers.

Mechatronic systems often decentralise the functions with the use of microprocessors to perform specified functions around a sub-system such as the combustion management in a car engine. This decentralisation brings in significant changes in the overall design, performance and maintenance of the product. With the introduction of microprocessors in a product new opportunities become available to designers to bring forward new products to the market which would not have been possible in the conventional way. In manufacturing process design, a mechatronics approach enables the use of smaller batch sizes with consistent quality.

As a consequence of these characteristics of a mechatronic system the product designer should look at the functional and other requirements of the product more objectively to identify sub-systems which could be controlled by a local microprocessor and simplify and enhance the product.

2.2 REQUIREMENTS OF A MECHATRONICS DESIGN SYSTEM

A mechatronics design system will be expected to provide a structure that can enable the integration of various technologies, their establishment and evaluation. The structure should enable the performance of the following procedures:
(i) Systematically recording the requirements made on the product.
(ii) Categorising them into several levels so that grouping them functionally is facilitated
(iii) Evaluating them in their order of priority
(iv) Establishing the corresponding design requirements
(v) Recording the proposed solutions (designs) that meet the design requirements
(vi) Analysing these candidate designs
(vii) Applying various scientific and engineering principles to optimise the design
(viii) Selecting the best design in the given circumstances

Design Function Deployment (DFD) developed at the City University Engineering Design Centre meets these requirements and is a suitable platform for cross-disciplinary product development.

2.3 OPTIMISING CRITERIA

In general a design should be optimised for "Quality". This is difficult to measure or define precisely, but will contain elements of cost, robustness and any specific requirements such as power/weight ratio. Robustness involves considering the effects of tolerances on component values and variations in materials and the production process. In mechatronics applications safety and reliability are important optimising parameters.

3.0 DESIGN FUNCTION DEPLOYMENT

Design Function Deployment, a formal system to be developed as a computer based design system, can be defined as a systematic method for the translation of explicit and implicit customer requirements as well as designer intentions into measurable design requirements (functions).

The system preserves traceability to the original customer requirements throughout the design, manufacture and use stages in the product or process. DFD enables the proper conception of a product at the design stage for manufacture and use. Within DFD the designer is presented with the opportunity to consider and resolve early (upstream) in the design cycle conflicting design requirements and constraints with other associated issues such as manufacturing, cost, maintainability, reliability, tooling, market and sales considerations etc, as well as the consideration of the integration of the mechatronics approach during the design process, in a simultaneous manner.

The development of a product starts with the identification of a need. This involves market research and from which the derived results are formally identified as customer requirements. These customer requirements are then translated into design specifications which can contain several levels and degrees of importance. Once the design specification is established the process of design synthesis begins. It involves imagination and original thinking in the design of the system and sub-systems. This search for different solutions is facilitated by DFD and results in many different possible solutions. The solutions are then analysed and the preferred solution is selected. The critical parts, the materials and manufacturing processes, the cost, reliability, robustness and failure mode analysis are some of the issues which are considered during analysis. Prototypes could then be produced to confirm the agreement of performance and cost with the design targets.

3.1 FEATURES

The DFD method is based on the charting technique QFD [3], which employs a series of 'what and 'how' charts to ensure a systematic approach to product development. It has five levels namely:

(i) Customer and design requirements level
(ii) Product characteristics level
(iii) Parts and component design level
(iv) Materials and manufacturing process design level
(v) Production planning level

To establish a product through these five levels much scientific and engineering knowledge and necessary software has to be utilised. In the software implementation of DFD, these programs are said to be in software level 2 while the creation of the five levels of product development are kept at software level 1. All the databases and rule bases are constituted within software level 3.

Figure 1 shows the schematic structure of the DFD system. The supporting modules in the software level 2 are developed as stand alone modules with a common data structure to enable the linking of these modules to the over-arching software at level 1.

4.0 SPEED CONTROL OF A DC MOTOR - AN EXAMPLE APPLICATION

A typical mechatronics system could involve a combination of analogue electronics, microprocessor, some software, a DC motor and a mechanical load. Figure 2 shows the block diagram of such an open loop motor control system. Cost constraints and initial design specification prevent the use of a tacho feedback system. The software will continuously sample the joystick position and update the motor control voltage. Much better results can be obtained if the software can compensate for the mechanical load, inertia and friction, and the DC motor characteristics. Ideally the transfer function of the motor and load are known, so the software can then compensate exactly.

Failure Modes and Effects Analysis should be used to ensure that the control system is fail safe. Design software can be used to model the analogue electronics and determine the consequences of each component failing. Robust circuit design should then be used to determine which components in the circuit are critical and require close tolerance components. The technique involves making a mathematical model of the circuit, inserting component values, predicting the performance, then altering component values within a given tolerance and repeating the calculations. This method is useful in failure mode and effect analysis, since component failures can be simulated by zero or infinite component values.

In practice there may be considerable variation in DC motor characteristics, even within samples of the same batch. The motor torque constant, back emf constant, and armature resistance all may vary. In the general case numerical values for mechanical inertia and friction are not known. Lack of data means it is often difficult to write software to compensate exactly for motor and load characteristics. Ideally the software should be able to compensate for wide tolerances in motor and load parameters.

Figure 3 shows a series resistor used to measure actual motor current. This can be used to estimate the motor speed assuming $v_m = v_b + iR_a$, where v_m is the voltage across the motor (known), v_b = back emf = $k\omega$, and R_a = armature resistance. The armature current is measured, so assuming numerical values for back emf constant k and armature resistance R_a, the angular velocity ω can be calculated.

The software knows the theoretical transfer function of the motor and load. The controller is initially programmed with reasonable, but assumed, numerical values for load inertia friction and motor constants. A known control voltage is applied to the motor and the actual motor current is measured. This is compared with predicted motor current and any large discrepancy is assumed to be due to inaccurate numerical values for motor and load characteristics.

The software could then change its assumed values thus effectively performing an experiment to measure motor and load characteristics. The problem is knowing which parameter (motor back emf constant, motor torque constant, armature resistance, inertia and friction) to change and by how much, since they all affect motor velocity. The Robust Engineering Design module within the DFD provides a formal method, based on Taguchi, of determining which parameter affects the output, their relative sensitivities and any interactions between them.

4.1 ROBUST ENGINEERING DESIGN AND TAGUCHI METHODS

Robust Engineering Design is a process that can make a product/process performance insensitive to variation originating from various sources during the life cycle of a product [4]. The term robust in this context means the robustness against variation. Taguchi methods apply experimental design and analysis to make the product robust. The method essentially has three steps: (i) Design of the experiment (ii) Conducting the experiment and making observations (iii) Analysing the observations and interpreting the results.

The design of the experiment depends on (a) the number of factors to be studied (b) the number of levels to be studied in each factor and (c) the interactions to be studied. The number of levels to be studied depends on the relationship (linear, quadratic etc) the factor is having with the performance characteristic involved. For instance to investigate a linear effect a two level study is enough wheras a quadratic effect would require three levels. Once the number of factors and the number of levels in each of them are decided, the design of the experiment can be represented in a matrix. The rows of the matrix represent trials and the columns represent factors. Taguchi recommends different orthogonal arrays from which to select these designs. The process of design when using these arrays reduce to selecting the appropriate columns to represent various factors to suit a given interaction pattern. To deal with interactions Taguchi provides a triangular table. A complete treatise on this is given in Logothetis and Wynn [4].

For the example motor speed control system the response is speed and the factors affecting speed are (i) Mechanical Inertia (ii) Friction and viscous damping (iii) Motor torque constant (iv) Motor back emf constant and (v) Motor armature resistance. Assuming no interactions and two levels per factor the design could be picked from the OA_8 from the Taguchi arrays.

In practice, the actual motor and load characteristics are not known, but high and low limits for each parameter can be set. With two levels, (high or low) for each of the five unknown characteristics, the speed characteristics could be studied. For an exhaustive combination, 32 trials should be conducted, but with Taguchi methods and use of OA_8 arrays, the design of the experiment could be conducted with 8 trials as shown below.

$$\begin{bmatrix} 1 & 1 & 1 & 1 & 1 \\ 1 & 1 & 1 & 2 & 2 \\ 1 & 2 & 2 & 1 & 1 \\ 1 & 2 & 2 & 2 & 2 \\ 2 & 1 & 2 & 1 & 2 \\ 2 & 1 & 2 & 2 & 1 \\ 2 & 2 & 1 & 1 & 2 \\ 2 & 2 & 1 & 2 & 1 \end{bmatrix}$$

The analysis part of the method utilises the technique called Analysis of Variance (ANOVA). ANOVA is a technique used for splitting the total sum of squares into contributions made by different factors. For a situation where there are five factors A, B, C, D, and E, the total sum of squares could be accounted for by the following equation:

$T_{SS} = CF + SS_A + SS_B + SS_C + SS_D + SS_E + SS_{error}$

Depending on the contribution of each factor, their inclusion could be decided in the final consideration. The factors with the most significant contributions are then considered to be of utmost importance.

4.2 ADAPTIVE CONTROL SYSTEM

Figure 4 shows a block diagram of a micro processor based adaptive speed controller. If the motor is not running fast enough more torque is required, which means more current. Processor No 1 effectively forms a transconductance amplifier since the actual voltage fed to the motor is calculated as the difference between, *measured motor current × amplifier gain* and *required velocity voltage*. The gain is effectively dependent on the motor torque constant. This will compensate for variations in mechanical load and to some extent variations in friction and damping.

The actual motor speed ω can be calculated from $\omega = (v_m - iR_a)k$. Assumptions are made about numerical values for armature resistance and back emf constant. Rather than write these numbers into the program they can be read into the program external registers as input. It is then possible for the second processor to change these values to suit different motors.

The joystick is interfaced to the second processor which will calculate an acceptable acceleration/deceleration strategy based on assumed mechanical inertia, friction and motor characteristics. The actual speed is then compared with required velocity and, if necessary, the motor constants are changed to suit. This then becomes a non-linear control system and there are implications for system stability. The first processor should have a fast response time and the second processor should respond to the joystick in a measured way and change the motor constants much more slowly.

The second processor performs an experiment to "Measure" the motor characteristics. Ideally, as part of the design process, data would be available on average, maximum and minimum values for back emf and torque constants. For a simple experiment, just two values of each constant are considered.

The software starts off with average values, compares actual speed with required performance and slowly alters the motor constants. From the Robust Engineering Design analysis in section 4.2 above, the most influential parameter can be determined. It is then possible for the controller to discover which combination of motor and load is in use.

By using two micro-processors, response time problems can be solved and diagnostic self-testing is possible, increasing system reliability. An alternative architecture, using fewer components, would be to run the transconductance amplifier loop as an interrupt service routine, and the joystick loop as a main program. Software counters and time delays in the main program can be optimised for relative time spent in the two loops.

4.3 DESIGN OPTIONS

DFD provides a method of determining and selecting between different design options. In this example, the major options are
(i) Number of processors - one with interrupt service routine or two with dual ported registers
(ii) Strategy for acceleration/deceleration
(iii) Relative execution speeds for the two loops
 - software execution times and the clock speeds
(iv) Time taken to change numerical values, and which parameters to change
(v) Different power amplifier circuits.

Selection criteria are (a) manufacturing cost (b) reliability (c) fail safe and (d) system performance.

5.0 CONCLUSIONS

The application of formal design methods can improve the quality and scope of cross-disciplinary products. DFD facilitates the designer to achieve this by providing the platform for a systematic integration of customer and design requirements and scientific and engineering expertise, in the product development process.

6.0 REFERENCES

(1) **BRADLEY, D. A. et al**, " Mechatronics - Electronics in Product and Processes ", 1st Edition, Published by Chapman & Hall, UK, 1991.

(2) **RIETDIJK, J. A.**, " Ten Propositions on Mechatronics ", Mechatronics in Products and Manufacturing, Lancaster University, September 1989, pp 2 - 3.

(3) **JEBB A et al**, " Design Function Deployment ", City University EDC Technical Report No 44.

(4) **LOGOTHETIS, N & WYNN, H. P.**, " Quality Through Design ", 1st Edition, Published by Oxford University Press, USA, 1989.

ACKNOWLEDGEMENTS

This work is supported by the Science and Engineering Research Council (SERC), UK

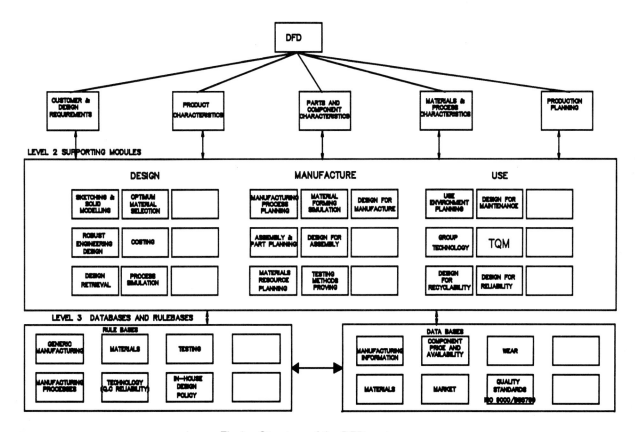

Fig 1 Structure of the DFD system

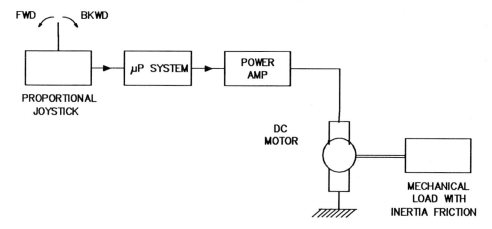

Fig 2 Open loop velocity control

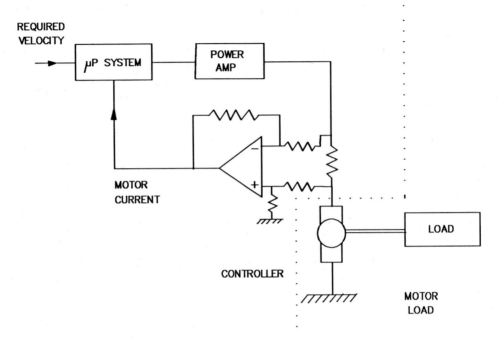

Fig 3   Measuring motor velocity

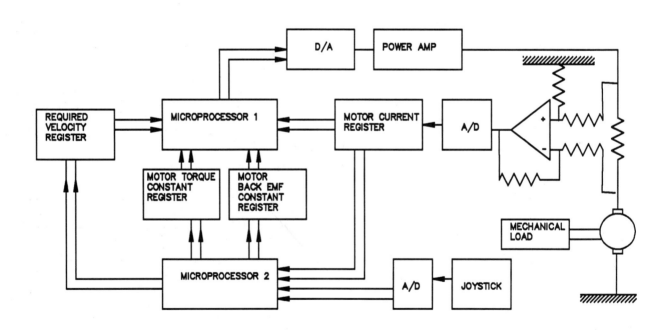

Fig 4   Adaptive motor speed controller

Development of an educational kit for mechatronics design

G R KISS, BSc, Dip Eng, **G S BELLIS**, BEng
The Open University, UK.

SYNOPSIS

A practical project for mechatronics education and training has been partially developed as a feasibility study. We report on the ideas that underlie the project design and on the progress we have made in developing educational materials for it in hardware and software. The materials are suitable for further development into a free-standing computer-based training module, or for incorporation into conventional educational courses as a practical project component. The materials also have relevance to rapid prototyping kits for mechatronics. Our motivation for this work is partly the overlap between mechatronics educational tools and mechatronics product development tools. A distinctive feature of our approach is the adoption of intelligent agents as a framework to deploying artificial intelligence in mechatronics. This brief paper concentrates on what we regard as novel in our approach and leaves out some aspects of our design like assessment, student activity timetable and the description of more obvious supporting student activities surrounding the central design exercise.

1. INTRODUCTION

Mechatronics is a very new subject area. There are several interpretations offered in the literature of what it comprises. In this section we briefly state our own prejudices as to what mechatronics is and then briefly state our educational objectives for developing kits for practical projects. Due to space limitations in this conference paper, this section has been very much abbreviated. A fuller discussion can be found in the complete paper, available from the authors.

We believe that *mechatronics is an evolving approach to engineering design*. Some of its distinctive features appear to be:

- The increasing range of different technologies that the designer has at his disposal, especially all branches of information technology, including artificial intelligence.
- The emergence of radically new design alternatives based on these new technologies as they rapidly mature and become cost-effective.
- The need for inter-disciplinary, integrative and systems-oriented approaches to design.
- The increasing complexity of the objects being designed.
- The trend towards rapid prototyping, incremental development and optimisation.
- The increasing rate of change in the technologies.

These features make the mechatronics designer's task more difficult than that of a more specialised designer. The mechatronic approach to design is a response to these difficulties and attempts to make them tractable.

If it is the case that mechatronics is an approach to design, then the educational objectives need to address the issue of teaching design in the mechatronics "style". We briefly summarise the objectives:

- To teach the student some practical aspects of *the mechatronics design approach*. Briefly, this is a structured approach to the improvement of some aspect of a machine's design, using integration and tradeoff between various technologies. As a result of resource limitations, creativity in generating possible solutions, will have to be limited to a few alternatives, but these should still offer a possibility for principled design decision-making.

- To teach the skill of analysing an existing system from the mechatronic systems point of view, i.e. to assess the contribution of the different technologies to the overall functionality of the system. This is important, since mechatronics is at present often deployed by re-designing an existing system through tradeoffs between technologies.

- To offer a practical exercise in the application and integration of the concepts and theories encountered in textbooks and other sources.

- To offer an opportunity to work with a few examples of component sub-systems that enter into the integration of a mechatronic product.

- To offer a practical opportunity to use some of the computer based tools and devices that are available to a mechatronics designer to aid in the design task. These are essential aids in coping with the complexity problem in mechatronics.

2. MECHATRONICS CONTENT; THE AGV DESIGN CONTEXT

The project content is provided by the mechatronics issues that arise in the *design of a miniature AGV system*, that is, a small vehicle with various sensors and actuators, controlled by one or more small computers.

A high-level AGV task is used as the context for a design exercise which itself tackles only a lower-level design issue. The high level task we have taken as an initial example is a restricted form of the *self-parking problem* for a vehicle. The project activities thus revolve around the problems of making the AGV capable of parking itself between two stationary objects, next to a kerb (see Figure 1).

The student's design activities are aimed at resolving a set of sub-problems involved in providing this functionality. Our main concern is to guide the student through a range of design exercises in which higher and higher quality performance is obtained against increasing financial investment, complexity and effort. We take an appreciation of such tradeoffs to be essential in mechatronics design education. Initially we have chosen to work with proximity sensors and simple on-off motor control.

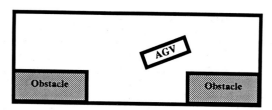

Fig 1   The AGV and its environment

One might argue that the self-parking task is not realistically supported by proximity sensing alone and that distance sensing should also be used. Our answer is that this is indeed the case, and future evolution of the project should move in that direction. However, a lot can be learned from attempting the task with proximity sensors only and finding out what performance limitations will result.

We have selected the use of microswitches and metal detector devices as proximity sensors as the target of the initial student exercises. Higher cost distance sensors may be added later. However, in terms of information flows and interactions between objects, these simple sensors form a sound and useful foundation to the study of the more sophisticated ones and to the more complex object interactions we might wish to deal with.

Thus, the student activities to be discussed below are concerned with making design decisions about the proximity sensors and about some of the software components that are needed to generate at least elements of the self-parking behaviour.

The impact of these design decisions on the high-level AGV action and planning functionality will be analysed only conceptually by the student, supported by computer simulation models. This will enable the student to experience at first hand the intricate nature of the mechatronic design process. *The student will be actively involved in considering the tradeoffs between technologies during the evaluation of his/her design decisions.*

2.1. Heuristics for self-parking with proximity sensors only.

In this paper we shall discuss a project scenario based solely on *proximity sensing* (informally defined in terms of the ratio between the size of the AGV and the distance at which some effect is being sensed.)

Thus the AGV has no sensory information about its position and orientation coordinates in its environment. It can only detect contact with the boundaries of the roughly T shaped free workspace. (The lack of position information makes this setup somewhat unrepresentative from the point of view of the standard AGV navigation problem, where position information combined with vision or sonar is the common practice.)

From the kinematic point of view, the prototype AGV is a non-holonomically constrained device. The drive motors are DC reversible motors, operated through relays. The two central facts that follow from these properties are that it can only execute the self-parking task by doing obstacle-following motions and that it can only move sideways by doing a sequence of translational and rotational movements.

These constraints however still allow *simplified versions* of the self-parking task to be solved. One such version is obtained by allowing only a subset of all possible starting configurations of the AGV, those that are defined by the AGV position being outside the "parking bay" area and the AGV orientation being at an angle less than 90 degrees to the "kerb-line", with its front pointing left in the diagram. In other words, the AGV will not have to reverse its orientation during the task.

We assume that the AGV knows (implicitly or explicitly) the following:

- The topological shape of its environment, (straight line boundaries, 90 degree angles occurring in a particular sequence).
- That some of the corners are convex, the others concave.
- The constraints on its permissible starting configurations.
- That its own shape is rectangular.
- The topological location of its sensors on its own boundary, i.e. left-front, right-rear, etc.

Initially it does not have any metric information about the dimensions of any object.

2.1.1. High-level actions

In order to solve the problem, the AGV needs to have some *high-level capabilities* like being able to recognise convex corners or being able to align itself parallel to a boundary.

A high-level task plan can be formulated in terms of such high level actions. This might involve moving to a boundary, aligning with it, and then navigating to the goal position. Details of this navigation process are described in the full paper.

Plans of this kind could be expressed in the form of rules or in various robot control languages. Both will be illustrated in the project handbook.

The project kit software provides some but not all of these high-level capabilities. The student will implement at least one by programming, *but in the context of other design decisions about the electronics and mechanics of the AGV system as a whole. A good candidate is corner finding and recognition.*

2.1.2. Low-level actions

The high-level actions mentioned in the previous section need low-level implementation in terms of simpler action sequences. Such a hierarchical plan decomposition needs to be followed down to the level of turning motors on and off and reading sensor states.

Take the concave corner recognition action as an example.

A heuristic can be derived from the fact that the AGV knows that its own shape is rectangular. Concave corners can be recognised in the environment by doing pattern recognition: detecting congruence between a convex corner of the AGV and a concave corner of the environment (see Figure 2).

Once in this configuration, small translational and rotational movements result in well-defined sensor contact patterns that can be used to confirm the recognition. *These patterns of switch closures are representations the AGV has of the environmental topology.*

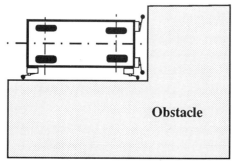

Fig 2   Concave corner recognition by shape matching

A preliminary to doing corner recognition is the ability of lining up one side of the AGV approximately parallel to a boundary surface. The algorithm for doing this is shown, expressed in the LabView visual dataflow language, in Figure 8 at the end of the paper which illustrates the flavour of the language and also shows how a knowledge-based analysis can be attached to locations of the dataflow diagram. We call such modules "agents", since they have goals which they pursue using the knowledge they have about the environment.

3. STAGES OF THE PROJECT

Space limitations in this paper preclude a comprehensive description of details. A fuller unpublished technical report is available from the authors. A brief account follows.

3.1. Introductory exploration and exercises

The purpose of these is to familiarise the student with the hardware and software setup and to "get the feel" of the overall system.

Included is a manually controlled "drive by instruments" exercise in which the student carries out the self-parking manouvre using a software front panel of controls and indicators on the computer screen. This gives an excellent opportunity for the student to experience at first hand the kind of perceptual/cognitive, planning and motor control problems that arise in the self parking task for an AGV that knows certain facts about its environment, but has only very limited access to that environment through the sensors. The software front panel has been implemented in LabView.

3.2. More detailed and partly quantitative analysis of the AGV system

In this activity the student will assess the characteristics of the total AGV system, comprising the AGV hardware, the AGV environment, and the control computer. This assessment is done in order to identify weaknesses in functionality with respect to performance in the target task, i.e. self-parking, that are potential targets for mechatronic design improvements. The assessment involves *qualitative and quantitative tests, measurements and inspection*. A major component of this activity is to assess features like accuracy of movement, repeatability, etc. The activity of analysis will be supported by the AGV system as a whole, including the use of the control computer to offer flexibility.

3.3. Learning to program the AGV

The AGV is programmed using a visual (icon-based) object-oriented language. This is the LabView system of National Instruments. Even though this kind of language is relatively simple to use, there has to be a stage during which the student learns to use it in an orderly and taught way. This is in order to reduce the duration of the learning stage, which, experience with the language shows, can take up to 50 hours for reasonable fluency in *general* programming. We do not intend to bring students to this level, but simply enable them through directed teaching, to do the *specific* programming tasks required.

3.4. The design exercise

Some of the proximity sensors are provided at present in the form of microswitches on the prototype AGV. These switches are themselves reasonably complex electro-mechanical objects and it is possible to give a mechatronic analysis of them. This will be part of the conceptual contents of the project.

A mechatronic analysis involves looking at an artifact and considering its functions from the point of view of all relevant technologies.

In the case of the switch some of the relevant technologies are *mechanical, electrical* and *informational*. The last is possibly the most important, since the device is used as an information gathering sensory interface from the AGV to its environment.

The mechanical and electrical design of the switch influences its informational characteristics. For example, the temporal uncertainty of event detection is dependent on the switch bounce duration and the electrical signal that can be derived during the bounce period. The spatial uncertainty of force detection is determined by the mechanical lever+roller+spring+trigger configuration in the switch design. The uncertainty results in a region of a vector space in which forces can be detected, while the sensor is blind to other regions. This is illustrated in Figure 3:

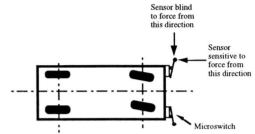

Fig 3   Directional sensitivity of microswitch

Sensor damage from forces outside some limit also needs to be considered in relation to the informational function.

The temporal uncertainty of switch closure results in an interval of time. The information that can be delivered by the switch as a sensor is the delimitation of a spatio-temporal zone.

The student is able to influence this zone by using more switches and by altering the mechanical characteristics and the mounting of the switch. One example, motivated by the desire to reduce the dead zone, is shown in Figure 4 below:

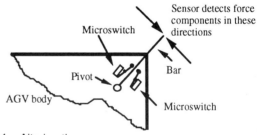

Fig 4   Altering the sensor range

A suitable design kit is provided for the student to use in implementing his/her design decision. The decision has to be made by the student in the context of the self-parking task, and one of its lower-level component actions (e.g. corner finding and recognition) that is selected for implementation.

The student will soon discover during exploration of the capabilities resulting from using microswitch sensors that with many designs undetected collisions could occur. On their own, a small number of microswitch sensors cannot support the self-parking task in the given environment. This leads to a realisation that there are separable requirements for detecting unexpected collisions with obstacles on the one hand, and using a sensor with which to actively explore and sample the geometry of the environment.

We believe that there is great educational benefit to be gained from a clear understanding of the action of sensors as information gathering devices; the cross-correlation of information from sensors of different characteristics (and so-called data-fusion); and the analysis and understanding of sensor arrays. An important consideration is the *spatial* sampling nature of such arrays in the context of spatial frequencies of the AGV environment and of the AGV itself.

There are also useful design considerations to be addressed in the context of the *temporal* sampling by sensors and the conclusions which may be drawn about the temporal nature (i.e. 'permanency' or otherwise) of the environmental objects (or other phenomena contributing to the 'transaction').

Consider, for example, the micro-switch sensor array illustrated in Figure 5, which has been developed as one design solution to detecting forces acting from any direction in a horizontal plane.

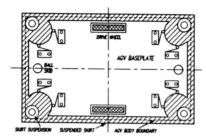

Fig 5   Omni-directional force detector

Here, an incident force (in our case a reaction to the probing force exerted by the AGV itself) is resolved along certain axes according to the location and orientation of the sensors. Force components exceeding the switch threshold result in a '1' output and the array as a whole thus generates a characteristic 8-bit output with 256 possible states. There is not of course a 1:1 mapping between incident force and the code generated - ambiguity exists because there is an 'envelope' of possible force vectors.

The student's design kit also contains photoelectric and magnetic proximity sensors that can be used in addition to or instead of switches. The student's task is to determine the optimal combination of sensors in the context of a broad range of design considerations. This leads to the evaluation stage of the project.

3.5. Evaluating the design

The student will evaluate the solution adopted with respect to a number of stated design criteria. Examples include hardware cost, software cost, accuracy, real-time speed, energy use, generalisability, etc. This activity is supported with the provision of technical information about the software and hardware components; tools for making measurements and observations and analysing their results; spreadsheet tools for asking "what if" style questions, and optimisation by mathematical programming; and tools for doing mathematical analysis.

3.6. Iterative improvement of design

As far as time allows, the student will attempt several cycles of design, implementation and evaluation on the same problem.

4. OUR PROGRESS WITH EDUCATIONAL MATERIALS

In this section we report on the progress with developing prototype materials.

4.1. Hardware

4.1.1. Hardware system overview

The current hardware system currently comprises four main items :-

(a) the model AGV

(b) the AGV environment (a controlled world to in which to investigate the AGV)

(c) the power supply/battery and PC interfacing box (might later contain processing power which controls any active environment and/or objects).

(d) a personal computer - developmental hardware has been exercised and tested using a DOS-based PC, while prototype software has been developed and is running on a MAC IIci.

The prototype AGV is a plastic frame with motors, sensors and electronic drive/interfacing modules (see photo).

The AGV is interfaced to the controlling computer via parallel I/O - via the parallel (printer) port for a DOS machine, via a plug-in I/O (Nu-Bus) board in the case of the MAC IIci. The parallel interface module at the AGV is partitioned from the AGV 'proper' so that it may be replaced by a serial interface (initially copper, later perhaps optical) with little difficulty. On-board processors will also be possible.

The AGV environment is a rectangular frame or 'cage'. It is used as a controlled environment and as a coordinate reference system. It may contain active and passive objects with which the AGV interacts in a variety of ways. The analysis and control of such interactions is a major educational tool for mechatronics design.

4.1.2. Hardware development strategy

For the reasons we outlined in Section 1 concerning design complexity and its management in mechatronics, the strategy of hardware development we have adopted is to evolve a set of functional 'black-box' modules which collectively would form a fairly flexible hardware architecture kit.

The design exercise would focus primarily upon system-level design and the decomposition of requirements into sub-system performance requirements, yet allowing particular 'black-box' modules to be 'opened up', re-designed, extended, replaced, expanded, where new performance characteristics or levels were appropriate or required.

We explicitly include also the philosophy of evolving a distributed processing approach to the deployment of microprocessors. This is in line with our approach to modularisation based on intelligent agents.

The AGV development is following this modular path. The set of 'black boxes' includes drive modules, steering modules, sensor modules, etc., with various performance characteristics, in addition to the more familiar electronic modules. Whilst an existing printed circuit carrier frame was initially adopted as a chassis (for speed and economy) it is envisaged that a more generic chassis structure with more appropriate morphological characteristics and with definable performance will be developed as yet another 'module'.

4.1.3. The AGV

The AGV in development exists in several different forms :-

(a) hardware development platform

(b) software development platform

(c) pre-production prototype format.

These are illustrated in the photograph and Figure 7 at the end of the paper. The hardware platform is a flexible bus-based arrangement designed to act as a test-bed for new hardware modules and exercised under test software. Phase 1 hardware modules were 'frozen' (in the design sense) and combined for small batch manufacture to form a phase 1 software development platform.

Development and evolution of additional modules (or sub-systems) is in hand as resources allow. The current emphasis is placed on extending the range of sensor sub-systems available.

4.2. Software

The microcomputer, due to its general-purpose nature and flexibility, is a vital component of the project kit. It has to support a variety of student activities, including control, observation, measurement, learning, simulation, writing, calculations, etc. It should, as far as possible, support computer-aided education by up to date means like hypertext, animation, models, mathematical aids, etc.

The main software tools that we may potentially use are shown in Figure 6.

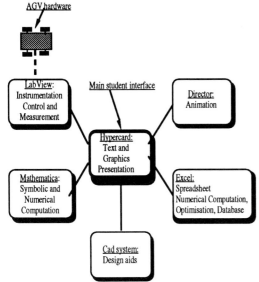

Fig 6 The main software tools

Of these, we have actually used Hypercard, LabView and Director to construct prototype subsystems.

4.2.1. Student interface

The student interface is concerned with the transfer of information between the student and the teaching system. It offers an appropriate GUI-based interaction style. It supports all the teaching functions, including textual, diagrammatic and animated presentation of teaching material. The student interface also makes it possible for the student to modify the AGV software system with ease and modest expenditure of time. It is easy to learn to use, self-documented, and satisfies modern requirements for HCI standards.

Hypercard. The main student interface and text plus animation presentation system. Other facilities are invoked from within Hypercard.

Animation. Embedded within Hypercard, using the Macromind Director system. This facility is used to present material in conjunction with explanatory text where animation can be used to advantage. An example of this is the information collecting characteristics of proximity and distance sensors.

LabView. The main tool for control and measurement instrumentation. It is a visual programming language the student uses for programming the AGV in . LabView is also used to teach about AI knowledge representation and reasoning in the mechatronics context, which is further discussed below.

4.2.2. Modular AI software integration

The AI software components are provided in a form that can be readily integrated with the rest of the Hypercard software system and with the AGV driver LabView software. In order for the student to be able to explore different types of AI techniques in design, these AI techniques are encapsulated into modules (OOP objects) that the student can readily invoke through the user interface (i.e. by mouse pointing techniques). We envisage the provision of neural networks, symbolic and nonsymbolic planners, search techniques, etc., in such modularly packaged form. An example is an advanced robot motion planner from Stanford University, based on the use of heuristic search techniques guided by an artificial potential field. The program is written in C and is used in a teaching module with a simulated AGV in either Labview or in Macromind Director.

In the LabView system icons are used to represent and invoke software modules written in the C language. The user need not distinguish between modules written in C and modules written in LabView. The graphical data flow network can contain both kinds of objects in the same iconic format.

5. CONCLUSIONS AND FUTURE PLANS

We have described the educational philosophy for, and a partial implementation of a teaching package for mechatronics design practice. We envisage the use of such a package in advanced university teaching at postgraduate level and in advanced industrial training.

Our future plans include the systematic exploitation of the overlap between mechatronics educational tools and mechatronics product development tools on the one hand, and the overlap between research activity and the development of educational materials on the other. Our major emphasis in the latter is the development of mechatronics sub-systems as intelligent agents.

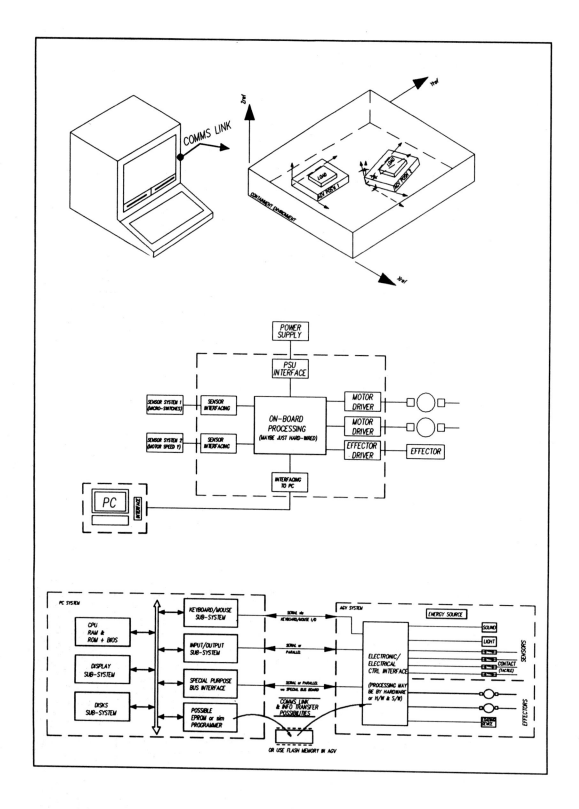

Fig 7 AGV system

Labview representation of part of an AGV agent whose goal is to align the AGV to a straight surface it believes to be close by on its left

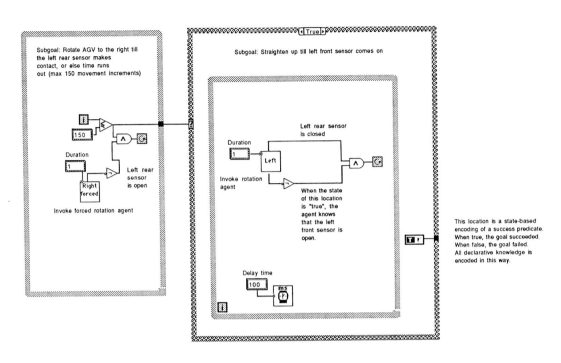

Fig 8   Example LabView program

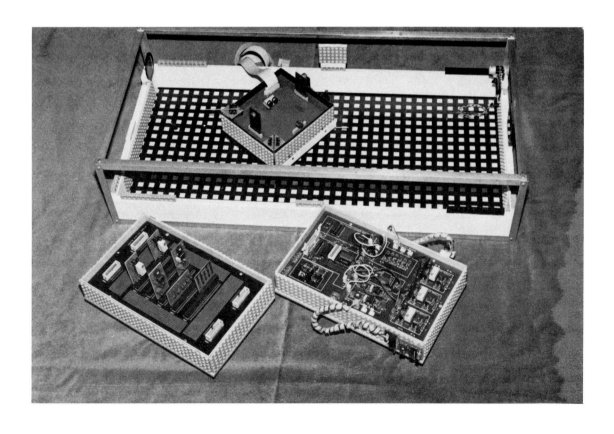

Fig 9   AGV hardware prototypes

Development of a new diagnostic system by Ford Motor Company and GenRad

R J SHORTER, BSc, AMIMechE, ACMA
Ford Motor Company Ltd, UK

SYNOPSIS This paper describes the development of FDS 2000 microprocessor based diagnostic equipment for use in Ford dealerships to diagnose faults in complex vehicle systems. Although launched for engine management systems, FDS 2000 is designed to address any microprocessor based electromechanical system, including new concepts not yet in production. The paper concentrates on the rationale behind the development, rather than technical details.

1 INTRODUCTION

Garage technicians have difficulty in diagnosing faults in these systems because they are a complex mixture of electronic and mechanical components. They often overlook mechanical problems that they would correctly identify in a conventional system because they are blinded by the apparent complexity of the electronic black boxes. Closed loop control means that there is often no clear relationship between symptoms and faults, and so the traditional reliance on training and experience is not effective.

A market led innovation was needed because of rapidly changing vehicle technology and the need to satisfy increasingly demanding customers. Market research had shown that:

o Users of everyday 'workhorse' vehicles demand very high reliability, while users of more exotic vehicles, such as a high performance, low volume sports car, will tolerate lower reliability.
o Throughout this spectrum of users, ALL users expect the dealer to be able to fix any problem at the first attempt.
o Customers find engine / transmission driveability problems particularly irritating as they occur continually and cannot be ignored.
o Consequently, there is a strongly adverse effect on a customer's re-purchase intention if he encounters a driveability problem which the dealer cannot fix at the first attempt.

Ford formed a cross-functional team with members from Product Development, Service and Computer Systems, to address the issues. It was felt that a complete strategy was needed for dealing with all aspects of mechatronics diagnosis, and that a cross functional team was the only way to focus sufficient attention on it.

2 DIAGNOSTIC STRATEGY

Ford's overall diagnostic strategy is:

o Maximise on-board self-test capability
o Provide serial communication links
o Develop off-board diagnostic equipment

Our in-house engine management systems already had a comprehensive self-test capability, and the Californian OBD II legislation pushes one towards on-board diagnostics. However, although it is possible to meet the requirements of OBD II - ie for the system itself to determine whether or not it is functioning correctly - it is easy to demonstrate that it is impossible for any practical on-board diagnostic system to diagnose a fault condition down to the level of detail needed for a garage technician to repair it. Far too much duplication of components and circuitry would be needed, with cost and weight penalties and - worst of all - additional unreliability.

But, a total reliance on off-board diagnostic equipment means too much intrusive testing with the risks of damage to connectors and consequent reliability problems. Also, closed loop control systems hold much useful diagnostic information internally, which is not available to an external tester unless a communications link into the system is available.

The optimum solution is a combination of the two. This which means that simultaneous engineering is a necessity; diagnostics must be considered earlier in the vehicle design process. It is no longer practical for Product Development to design it, Manufacturing to make it, and then Service to try to work out how to repair it. Complex systems need diagnostics engineered into them, and time has to be allocated for writing software for the off-board tester. This is another reason why a cross functional team was a necessity.

Although it is anticipated that implementation of the strategy will reduce warranty costs, this was not the main purpose. The principal objective is to increase customer satisfaction by eliminating unnecessary repeat repairs.

3 COMMERCIAL CONSIDERATIONS

In general, electronic systems are very reliable, even when subjected to the arduous conditions of use that a motor vehicle is subjected to. Coupled with the wide variety of different vehicle combinations that it is possible to build, and the low take of some of the optional systems, this means that even a technician working in a large, busy, dealership is unlikely to regularly encounter similar faults.

We have 8 700 dealers throughout Europe, and it is likely that most of them will never see - for example - a Cosworth 24 valve Scorpio with an engine management problem. This is because this model is built in relatively low numbers, sales are likely to be concentrated in certain geographical areas, and the system is very reliable. But, ideally, every one of those dealers should have a diagnosis capability in case a motorist breaks down and requires assistance. We can get spare parts to most dealers overnight, but he has to be able to diagnose the fault quickly and accurately on his own.

This means that a large, expensive, state-of-the-art diagnostic machine was not a practical proposition. We had to produce something that was affordable by all dealers - even a small, retail dealer in rural Spain.

Furthermore, a common problem was technicians overlooking mechanical faults that they would correctly identify in a conventional system because they were blinded by the apparent complexity of the electronic black boxes. This meant that the new diagnostic system had to be very user-friendly - so that it would always be used - and had to include sensible and relevant prompts to remind the technician to perform conventional checks where appropriate. This implies a good user interface and substantial memory.

By 1990 there was a high degree of urgency. In Southern European markets the use of electronic engine management systems - an optional feature - was very low, but for 1993 model year all petrol engined vehicles would have these, ahead of the EEC emissions regulations. The local technicians did not have the necessary experience or training. Furthermore, the more sophisticated European markets that were already familiar with engine management systems and anti-lock brakes would soon be receiving additional electronically controlled vehicle systems on new models.

This was our dilemma:

o We wanted something that was relatively cheap, but very capable
o No existing equipment came even close to meeting our requirements
o We had to produce it within a reasonable budget
o We needed it to be developed and launched in all European markets within 24 months

4 VENDOR SELECTION

We specified our requirements and invited tenders from all the vendors we could find that had proven experience in this field. It soon became clear that it was feasible to develop dedicated electronics in this timeframe, but that providing the software was going to be a problem. Most vendors suggested writing custom software in a standard language (usually C), but were unable to convince us that it was possible to start from scratch and achieve the degree of functionality and polished user interface that we were looking for in the available time.

GenRad UK are a subsidiary of a US company formed in 1904 that used to be called General Radio; their main business is large automated board testers for use by electronics manufacturers. We knew they had the hardware design and manufacturing capability, and they were able to demonstrate an LCD display with touch screen overlay which offered total flexibility in the user interface.

More importantly, they had provided the Jaguar Diagnostic System (JDS) (1) in 1985, built around their Universal Field Tester that used a Zilog Z80 processor. This had caused them to develop a high level diagnostic language called PAL, and to start constructing a dedicated software development environment. They were already working on PAL2, to be compatible with the Motorola 68000 microprocessor.

JDS was too expensive and too large for us. It was also obsolescent technology. But, GenRad had the software development tools that were necessary to meet our timing requirements and were able to design new portable hardware using the latest surface mount technology.

I cannot over-emphasize the importance of a formal, software development environment for this type of project. Without it, there was a choice of spending most of the time developing new hardware, leaving time for only a simple software implementation, or, use currently available hardware and spend most of the time on software development. The software development environment enabled us to develop hardware and software in parallel, evaluating alternative user interfaces and presentation style on a system emulator running on VAX stations, before the hardware was available.

5 FDS 2000 HARDWARE

FDS 2000 had to meet these requirements:

o Intuitive user interface
o Optimum ergonomics in what is an ergonomic minefield
o Unambiguous operation in eleven different languages
o Diagnose mechanical, electrical, and electronic faults

The heart of the system is the Portable Diagnostic Unit (PDU). This has:
- LCD touch screen, 120mm by 75mm, with graphics capability
- 68000 processor
- 1 Mb of memory
- comprehensive data acquisition hardware
- waveform generator
- communications ports for RS232, modified RS485, Ford DCL, and ISO 9141.

It can run on internal rechargeable batteries, or from the vehicle's electrical system.

Various probes plug into this unit for measuring: voltage, resistance, frequency, pulse width, duty cycle, etc; provision is made for current, temperature and pressure sensors in the future.

The pixel addressable graphics capability of the screen gives complete flexibility for diagrams, large numerals, and bar graphs, as well as three standard text fonts. In combination with the touch screen, it avoids the need to have permanently marked buttons, and allows the software to present to the technician keys that are relevant to the current situation, correctly marked, without any redundancy. Particular design effort was put into selecting a screen that is readable in the widest possible range of lighting conditions from complete darkness to bright sunlight.

Ergonomics was quickly defined as a major problem area. A case design for the PDU was eventually developed which incorporates hinged legs to enable it to lay flat, stand up like an easel, and hook onto a variety of locations in the under bonnet situation. The case is rubber covered and tough enough to stand up to constant use in a workshop. A carrying 'holster' is provided to hold the PDU together with cables and adaptors.

There is a Vehicle Interface Adaptor (VIA). This is for use in conjunction with the PDU on systems which do not have a serial connection, and for detailed pinpointing of some faults on serial link systems. It is basically a relay multiplexor which enables the PDU's measurement capabilities to be tee-ed into the multiway connector between an Electronic Control Module (ECM) and the vehicle's wiring harness for automated testing. It also contains isolating relays to interrupt certain control lines to the module, and current source and sink capabilities for driving actuators. Alternative cables are available for different vehicles systems.

Semiconductor multiplexors would have been smaller and cheaper, and adequate for data acquisition; these were recommended by most of the suppliers who tendered for this project. However, the use of miniature reed relays enables us to measure low impedances accurately and to route the waveform generation capabilities to the ECM and/or its various actuators.

A base station is provided for permanent mounting in the workshop. This provides storage for up to two PDUs and one VIA, with their "holsters", cables and adaptors, and battery charging. The PDUs' batteries are recharged whenever it is connected to the base station or a vehicle, and the charging rate is controlled automatically to maximise battery life.

The base station also contains a CD-ROM drive and the necessary electronics to enable diagnostics to be downloaded to the PDU. The enormous 640Mb capacity of this medium enables a complete set of diagnostics, with text in each of the 11 natural languages encountered in Europe, to be stored on a single disc. For batches of a few thousand, the discs are cheap to reproduce and they can be sent through the post without fear of damage.

Once the hardware has been manufactured and supplied to dealers, it would be impractical to update it, and so care has been taken to include all foreseeable measurement requirements. Future systems will be covered by new software with possibly the use of "fat" cables, containing protocol converters, where necessary.

6 FDS 2000 DIAGNOSTIC SOFTWARE

The fault population considered is the whole fault population for the system under investigation. All electrical faults are covered, electro-mechanical faults where possible, and guidance is given on purely mechanical faults. Interactive tests are used where part of the function can be observed electrically by the FDS 2000, but non-electrical measurements, observations, or adjustments are required.

Wherever possible, the technician is not expected to make decisions; diagnosis follows a logical progression down a "fault tree". However, prompts such as "Consider fuel contamination" inevitably require a judgement from the technician.

Back-probing of connectors is not permitted on Ford vehicles, hence the use of the VIA when it is necessary to take direct measurements on a live system. However, as little intrusion as possible is made into the system, and when it has a serial link it is usually possible to diagnose faults without disconnecting the ECM. Technicians are required to probe the front face of connectors that have only a few pins, but obviously this cannot be done while the system is functioning.

7 SUMMARY

The total diagnostic system of on-board self-test, serial communication link, and FDS 2000, is able to detect all hard faults. Many "in-range failures" of sensors are caught by the off-board diagnostic equipment, and most intermittent faults are trapped by the on-board self test.

REFERENCES

(1) Andrews, M.J. and Clarke, A.D. Development of a service support system for microprocessor-controlled vehicle electrical systems. Proceedings of the Institution of Mechanical Engineers, 1986, Vol 200 No D5, S43-S51.

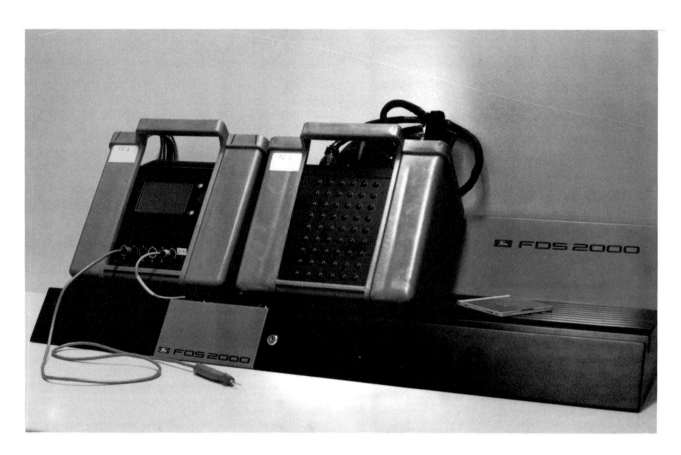

Fig 1   FDS 2000: complete system

Fig 2   FDS 2000: Portable diagnostic unit (PDU)

From woven bags to expert systems and broken digits

C J FRASER, BSc, PhD, CEng, FIMechE, MInstPet, J S MILNE, BSc, CEng, FIMechE, G M LOGAN, BSc
Dundee Institute of Technology, UK

SYNOPSIS: The paper describes the implementation of a multi-disciplinary mechatronics philosophy in the promotion of a number of consultancy and product development activities carried out in the department of mechanical engineering at the Dundee Institute of Technology. The examples given serve to illustrate the range, diversity and not least the benefits associated with the implementation of machine intelligence in the automation of the various mechanical functions.

1 INTRODUCTION

Within the last ten years there has been a proliferation of microprocessor technology in a wide ranging variety of applications to mechanical systems. In general this has provided the enhanced dimensions of intelligence and flexibility to the mechanical systems which incorporate the new technology. Typical mechatronic products are epitomised in the forms of the industrial robot, the modern video-recorder and the automated banking machines which have since become a regular feature of everyday life. These are all mechatronic products and the definitive characteristic which classifies them as being 'mechatronic', is that they all invoke the transfer of information either to, from, or within a mechanically based system employing micro-electronics technology. Information exchange is the key element in a mechatronic system and that which forms the basis of the system's 'intelligence'.

The products cited above as mechatronic systems are all very sophisticated, mechanically complex and are the end result of thousands of man-hours of development. Two of the examples which are described in this paper are much less mechanically complex but they nonetheless embody aspects of information exchange within their design. The attribute of information exchange within these simpler systems illustrates the added flexibility which has enabled the important fine tuning and optimisation functions to be carried out so readily. Thus it would appear that products which are designed around mechatronic principles from the outset are much more amenable to running adjustment and can apparently be adapted more easily to cater for an often variable product specification.

2 WOVEN BAGS

Dundee has had a long established tradition in the weaving trade which can be traced back to the earlier part of the last century. At one time renowned for the 'three J's', representing jute, jam and journalism, Dundee's former 'J', for jute, has all but disappeared from the commercial marketplace. The weaving industry however is still a major employer in the area, but the raw material of natural jute has been superceded with polypropylene.

The example described concerns a local company, AGRIPAC Ltd., who are a major manufacturer of woven polypropylene bags. These bags are designed to carry any granular product and are woven as a continuous tube in a rotary loom. The woven tube is subsequently cut to length, stitched together and a handle, formed from polypropylene cloth, attached to complete the finished product. Apart from the actual weaving process, the operation is essentially labour intensive and repetitive. The handle attachment process was particularly tedious and a project was sponsored to develop a partially automated system to perform the strength sapping routines in the handle attachment operation.

Initial studies were carried out on the existing manual operation in order to establish and perhaps modify the basic sequence of operations involved in forming a completed handle. Essentially an operator has to pull a suitable length of handle material from a roll of cloth, wrap the cloth round previously cut slots in the bag, grip the loose end of the cloth, tension the cloth from the roll end, cut the roll, traverse a sewing machine across the cut ends of the handle and finally remove the finished product from the machine. A gripping attachment and a cloth feeding mechanism are pneumatically operated but the rest of the sequence is performed manually. The basic remit of the project was to automate the entire sequence apart from the manual loading of the unfinished bags into the machine. For the interim prototype, it was decided to leave the automation of the sewing operation for a later stage in the machine's development.

The basic prototype was formulated round a system of pneumatically activated grippers and slideways, see figure 1. Lever action microswitches were used as end of stroke sensors on all cylinders. The prototype automatic machine however differed in basic operation to its manual counterpart in that the flow of handle material was constrained to move in the vertical, as opposed to the horizontal direction. The company had suggested that such a modification, which would of necessity rotate the orientation of the sewing machine through 90°, might reduce the downtime for sewing machine maintenance. Sequence control was implemented through a Mitsubishi programmable logic controller, or 'PLC'. The particular model selected provided a 40 line I/O capability with 24 input lines and 16 output lines. Extended I/O capability can be coupled in as an add-on unit giving up to a further 40 lines for input and output functions. For ease in programming, pilot operated five port spool valves with double acting solenoids were used to drive the various cylinders. The control program was written in ladder logic format using the MEDOC (1), development software. MEDOC constitutes a versatile and user friendly environment which greatly facilitates the writing, editing, storing and downloading of ladder logic programs suitable for running on Mitsubishi PLC's. An alternative method of programming is to write the actual mnemonics directly to the PLC via an input keypad. A listing of the required mnemonics for a complex sequence however may well extend over several pages of A4 and this method of programming is generally associated with a greater susceptibility to error. The MEDOC system however can also be enhanced with an I/O monitor which can be used as an aid in the debugging of test programs in simulation mode.

Initial problems with the operation of the automatic machine were related to the nature of the material being handled. Polypropylene cloth is all too flexible and the basic lessons learned from experience were that the material can never be completely released, nor must it be allowed to overhang from any gripper by more that about 30 millimetres. These problematic characteristics of the material necessitated some design changes and some minor resequencing of the original cycle but the machine has subsequently performed satisfactorily over numerous operating cycles. To complete the automation of the handle attachment process it remains to incorporate a metered flow pneumatic drive for the sewing machine. It is intended also to combine the cloth cutting operation with the sewing action in order to simplify the basic process and development trials are ongoing.

In the development of the automatic handle attachment machine, the use of an industrial PLC was envisioned from the very beginning of the project at the conception stage. For product development activities, the PLC easily accommodates alteration and modification. With a little experience, an entirely different sequence can be reprogrammed in minutes, or an existing program edited in a similar timescale. The mechatronic approach in this instance has resulted in a system which exhibits in-built flexibility obtained through the machine intelligence and information transfer facilities provided in the form of the microprocessor based controller.

3 EXPERT SYSTEMS

The terms 'expert system' conjure up many different interpretations to the many different users and developers of this technology. In the present context the expert system refers to a software based entity which operates in the basic role of a supervisor in a mechanical application. The target mechanical system is, in fact, an automatic teller machine, 'ATM', manufactured by NCR Ltd Dundee and the project is centred around the longer term development for the future generations of such machines.

The current project is an ongoing development from previous work undertaken on behalf of the company. In an ever competitive market, continual efforts are made to reduce the overall time spent on product development between the stages of mechanical assembly to field commissioning trials. For the particular case of the ATM, considerable time was being lost in waiting for appropriate software to control the machinery according to changes in the mechanical design. It was perceived that a suitable PC driven test facility would present the opportunity to carry out the mechanical debugging functions independently of the awaited system software and the basic remit of the original project was to develop a flexible stand alone test bed. The main project requirements were that the test facility software should be user friendly, be capable of being operated by engineers and technicians with limited computing ability and that the system should cater for the full range of inputs and outputs applicable to NCR based sensor and actuator products. Other specific requirements were the need to be able to save and edit programs and to log a retrievable record of the completed cycle count in the event of a system failure.

Using an actual ATM cash dispensing mechanism, a suitable interface was constructed to provide the communication links between the mechanism and the controlling computer, see McEvoy (2). The mechanism incorporates two basic forms of opto-electric sensors. These are encapsulated LED and photo-transistor pairs and separate LED transmitter/ photo-transistor receiver pairs. The range of basic actuators included various solenoid driven mechanisms, capacitor type ac motors and a solenoid based vacuum mechanism for picking up banknotes. Later model ATM mechanisms use stepper motors and dc motors in favour of the less flexible ac motor drives. The complete input/output capability is as follows :-

Outputs

6 x 24 volt dc solenoids
4 x 24 dc motors
4 x 110 volt ac motors, both uni-directional and bi-directional
4 x 24 volt dc switching signals
4 x 24 volt dc stepper motors

Inputs

6 x type 1 sensors, encapsulated LED/photo-transistor pairs
6 x type 2 sensors, LED transmitter/ photo-transistor receiver pairs

Digital input/output communication between an IBM-PC and the mechanism was provided through two separate 48 line I/O cards. The cards selected were the AMPLICON PC-14-AT model with one card dedicated to input functions and the other dedicated to outputs functions. For the output card, the port signals were first of all inverted and then used to control opto-isolated relays, which in turn would switch in the power supplies to the various solenoids and motors. Stepper motors were driven directly from NCR driver boards operating via port control signals. Input sensors required both a 12 volt and a 5 volt supply, the latter being used to drive the photo-transistor circuits which formed the basis of the input logic level signals.

The first level controlling software was developed using Borland's TURBO-BASIC and this was written in a hierarchical menu system where the user can page through the options until the desired functions are displayed. On selecting a function, the software then downloads the required CALL routines to a disc file which is being constructed within the program. The disc file, when completed, will constitute the mechanism controlling program and will be in the form of a series of subroutines followed by a series of CALL routines which will implement the selected machine sequence as prescribed by the user. Within the machine controlling program, the number of cycles completed is continuously updated and written to another disc file. This provides a non-volatile record of the number of machine cycles carried out.

Further software development, Forrest (3), has included a modular object oriented software rewrite in TURBO-C incorporating an icon based menu selection from a mouse. The updated software can display, edit, print and run the chosen sequence and additionally has the potential to record the system status at all points in any machine sequence. This provides the engineer with an improved maintenance tool in the event of a system crash. This particular version of the hardware/software system is currently in use at NCR's manufacturing plant in Dundee.

Following on from the success of this venture, the next logical progression was to further enhance and improve the intelligence capability of the system. To this end, NCR have fully funded a pilot study on the development of expert system technology to complement applications involving the ATM. Figure 2 shows a general view of the document handling sub-assembly of the ATM manufactured by NCR. The electro-mechanical complexity is self evident and the main problem in developing an effective supervisory expert system is related to time, or rather the lack of time. In general, the expert system monitoring phase would start when the leading edge of a note is detected. The system would then have to complete its monitoring cycle in less time than it takes for the note to pass under the sensor. With additional stress and wear monitoring functions in operation, the controlling program must be rapid in execution.

At the initial development stage it was proposed to simulate the flow of data to and from the mechanism in order to test the results of decisions made by the system. The intention here is to fine tune the control strategy to eliminate disruptive interactive and knock-on effects. Two data structures are envisioned where one will depict the current status of the system and the other will indicate the flow of media through the mechanism. In time critical parts of the cycle, global monitoring will be temporarily abandoned in favour of local monitoring with priority given to the key phases of the cycle. The monitoring processes will also be selectivly restricted to the more commonly occurring problematic areas of the mechanism. Currently the project has reached the stage where the system rules are being formulated and tested. An expert system shell, CRYSTAL (4), is used to enter the rules and other directives into a knowledge base and an inference engine, which interprets the knowledge base, is then applied to a custom made interface where the mechanism is simulated through I/O monitors. On completion of this phase of the project it is expected that a rigorously tested knowledge base will be available as a cornerstone for the development of the final version software.

Initial studies have indicated that the inclusion of expert system technology into an ATM may well necessitate some major redesign of the basic mechanisms. The biggest problem to be addressed however will be how to accommodate the additional checks performed by the software within preferably reduced cycle times for the ATM mechanism. It is critical that the time allocation for knowledge based decisions does not compromise the basic monitoring and control functions within the mechanism. The programming language currently proposed for this work is C^{++}, an object oriented language, which would allow the best techniques in software engineering to be used to evolve a system to an industry standard including a human/machine interface compatible with the new EC directive on software interfaces, see Bevan (5). The use of C^{++} will also enable the creation of the maximum number of re-usable modules and a library of control functions which could be used in a range of other related projects.

4 BROKEN DIGITS

The 'broken digits' cited in this third example refers to damaged human fingers and more specifically to the application of mechatronic principles in the development of a mechanical device which enhances the speed of the recovery process following such injuries.

During the 1970's, Salter (6), developed the system of Continuous Passive Motion, 'CPM', for the treatment of joint injuries. The underlying rationale is that passive motion generally reduces muscular atrophy and helps to keep the joint flexible. Therefore by minimising the period of immobilisation, the healing process is generally accelerated. This basic principle is now widely applied in the medical profession and a high priority is always placed on getting the patient moving as soon as possible after surgery. For joint surgery the aim is to keep the joint moving and to progressively increase the range of active motion. There are already in existence a number of systems available for implementing CPM for the hand, knee, shoulder and elbow. None of these systems however have addressed the questions of how much movement, or intensity of movement is optimum.

The prototype development system was based on an 8-bit single chip micro-controller, (the Intel 8052). This system enabled the closed loop control of both position and speed, and featured the measurement of applied force on a single actuator rod. Clinical trials have been carried out using the prototype system which was designed for use by orthopaedic department staff with little, or no, computer literacy. The controller was supplied with a series of compatible I/O interface cards which were used to provide the motor drive signals and the position feedback for the controlling algorithms. Set points are selected by activating the drive system through a number of manually operated switches and comfirming the set up through additional switches.

An upgrade system, see figure 3, is currently being developed using a standard IBM-PC equipped with digital I/O interface cards and analogue input facilities. This will enable a greater range of variables to be measured and stored which will include position, force, speed and joint angles for each of four individual finger actuators. The full range data acquisition system will be operated through the Microsoft WINDOWS™ environment and will provide the user with real time data graphs and a number of optional control strategies in addition to positional feedback.

The drive actuator system, which is mounted on the patient's fore arm, consists of a number of actuator rods driven by miniature dc servo motors. The linear translation is provided through a drive nut and lead screw arrangement. Mounted on the carrier nut is a strain gauge based force transducer which is configured to measure the applied force imparted to each drive rod. Positional feedback is obtained from a linear potentiometer. At present the potentiometer is attached directly to the actuator assembly but future enhancements will seek to incorporate the potentiometer as an integral part of the actuator structure. All drive and feedback signals are carried along an umbilical from the control unit to the actuator system. The patient has the ability, at all times, to stop the system running using an 'emergency stop' switch. This, in fact, only pauses the system which may be restarted at any subsequent time. Under normal operating conditions, the clinician may select continuous motion or alternate motion and rest periods. This feature has been added so that a quantitative investigation on tissue relaxation effects may be conducted in addition to the analysis of the applied forces and moments.

There are a number of commercial CPM systems available at present but none of them provide any method of collecting data on the effects of CPM during treatment. In addition none of the existing systems have the facility to vary the type of motion applied to the joint. The current system being developed will have the capability of applying programmed motion and force patterns to injured joints and will provide basic quantitative data on the applied kinematic forces and moments. Ultimately the objective is to shed some light on the bio-mechanical aspects of joint recovery. The example however provides a good illustration of the inherent flexibility associated with a mechanical system which has been built around mechatronic principles.

5 CONCLUSIONS

The three examples described provide a reasonable, if limited, illustration of the diversity of applications for mechatronic products. Although the projects are relatively straightforward, they all incorporate a systems approach to design which is consistent within a generalised mechatronics philosophy. The three projects can be described in completely general terms which can be categorised under the basic sub-headings of the mechanical functions, the machine/controller interface and the intelligence based controller.

The mechanical functions are centred around the mechanical power transmission in the various forms of levers, linkages, gearing, pulleys, pneumatic systems and electrical drives. Associated with the mechanical functions are the materials selection aspects for stress, cost and manufacturing considerations. These aspects are shared in common with the development of virtually any mechanical artifact.

The machine/controller interface for the controller input side, encompasses all of the sensing functions performed by the system. This includes the circuitry required for both signal conditioning and the connection between the analogue based mechanical system and the digital controller. The controller output side additionally features the digital to analogue conversion system and the power amplification circuitry for various motor, relay or solenoid drives. Although the machine/controller interface may be made up of many different sub-systems, the basic operational duty is to provide the analogue/digital and the digital/analogue links between the controller and the controlled mechanism. As in any design exercise, cost and fitness for purpose also figures prominently in the development of the interface between the machine and its controller.

Lastly there is the controller itself which can be one of many different forms. These may range from simple logic, or relay circuits, to an Application Specific Integrated Circuit, or ASIC. In increasing complexity, the controller may also be in the forms of a single chip, or single card microcontroller, or on up the embedded intelligence hierarchy to Programmable Logic Controllers and full PC based systems. The choice of controller will depend upon the complexity of the functions to be performed and the speed of operation required. This may have ramifications on the software selected for the purpose. Other considerations include the flexibility and adaptability required, the working environment of the controlled mechanism, reliability, maintenance aspects and costs. The machine/controller interface may also significantly influence the choice of controller to be used.

The common factor within the generalised system, and indeed that which makes the applications mechatronic, is the transfer of information. This is the basis of any machine's intelligence and the vehicle which promotes adaptability and flexibility at the product development stage. Intelligence may be enhanced by providing additional sensory functions and extended logical interpretation capabilities. In essence, the addition of sensory functions and logical attributes is an attempt to imitate the natural capabilities of human beings. In this respect, the higher the level of simulated intelligence within the system, the more versatile and flexible the system is likely to become. At the concept stage however it is important that the system is not over-designed with unnecessary intelligence. Economic constraints and manufacturing considerations should generally prevent this from happening.

The three examples cited in this paper have all benefitted greatly from having the attribute of intelligence incorporated within their designs. The greatest advantage derived has been the ease with which the modification and fine tuning functions have been actioned within the development stages of the prototypes. This has undoubtedly expedited the latter stage development activities in all three applications.

6 ACKNOWLEDGEMENTS

The authors wish to thank the Norman Fraser Design Trust for their support in the development of both the automated handle attachment machine and the continuous passive motion machine. British Council and Arthritis and Rheumatism Council Funding to support the CPM development program is also gratefully acknowledged. Thanks are also due to the industrial collaborators, AGRIPAC Ltd Dundee and NCR Ltd Dundee for their interest, financial support and encouragement in progressing the various projects.

7 REFERENCES

(1) MEDOC. Medoc - Manual, Programming and Documentation System for MELSEC PC Systems, Mitsubishi Electric UK Ltd., 1989.

(2) McEVOY, D. Computer Controlled Motion of a Mechatronic System, B Eng (Hons) project report, Dundee Institute of Technology, 1991.

(3) FORREST, S.J. Enhancement of Man/Machine Interface Software for use with an NCR Auto-Teller Machine, Post-graduate Diploma in Computer Based Engineering Systems, project report, Dundee Institute of Technology, 1991.

(4) CRYSTAL Intelligent Environments - CRYSTAL - version 3, IBM Corp, Microsoft Corp, 1987.

(5) BEVAN, N. Enforcement of HCI, Computer Bulletin, vol 3, part 4, 1991

(6) SALTER, R.B. Motion versus Rest: Why Immobilise Joints, Presidential Address, Canadian Orthopaedics Assoc., *Jour of Bone & Joint Surgery*, vol 64b, No 2, 1982.

Fig 1   Handle attachment machine for woven polymer bags

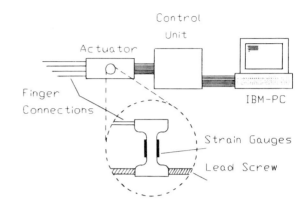

Fig 2   Document handling sub-assembly of an ATM

Fig 3   Continuous passive motion system

Integrating the cutting and sewing room of conventional garment manufacture through automated stripping

C A CZARNECKI, BSc, MPhil, **A PATERSON**, BEng, PhD, **B BRAMER**, PhD, MIEE
CIMTEX, Leicester Polytechnic, UK

SYNOPSIS Stripping is the process by which garment piece parts are removed from a multi-ply stack located on a cutting table and loaded onto a transportation system which feeds the sewing stations that perform the 'making up' operations. Research has identified manual stripping as a potential bottle-neck in garment manufacture aiming at quick response and low levels of work in progress. This paper describes the current state of progress of a programme of work aimed at automating this process and introduces a methodology for the intelligent scheduling of robots with uncertainties working in real time.

NOTATION

N   number of plies
Ta   time to automatically strip one garment piece
Ts   time to manually strip a stack

1 BACKGROUND

The emergence of developing countries as a large source of cheap labour, exporting to the west, has opened up a vast garment supply area, offering the production of garments at prices unattainable in the west. It has been estimated that the cost of producing garments in these countries is of the order of one tenth that of producing the same garment in Europe. Unless the western clothing industry meets the challenge effectively, it will lose, not only the clothing sector of the industry but there will be increasing pressure for the production of textiles to be located where the subsequent garment making-up industry is sited.

Areas to be addressed in order to prevent this happening are the improvement of quality and efficiency whilst reducing the labour intensity and susequent cost of the garment manufacturing process.

To a certain extent, the textile and clothing industries have mimicked the progress of the engineering industry in that islands of automation have been formed to produce some of the easier to handle sub assemblies. If Europe is to compete successfully in world markets as manufacturers they have no choice but to develop these systems and to continue to concentrate research efforts on improving manufacturing flexibility to respond quickly to retail demands.

CIMTEX is an £8 Million project which was established in 1989 to help the UK knitwear industry become more competitive. It involves collaboration between 38 equipment manufacturers and 5 academic institutions and is part funded by the Department of Trade and Industry (DTI). The emphasis is on Computer Integrated Manufacture in clothing but the application is to both Small to Medium sized Enterprises (SME's) as well as the larger organisations who are perceived to employ high technology solutions to production problems.

2 INTRODUCTION

In general, the single most important economic factor in garment production is the efficient use of the fabric from which the garment is made. In order to generate the shaped pieces to produce a garment, the roll of fabric has to have the piece parts cut from the fabric. The material between the cut pieces is waste. The amount of waste fabric is a very important economic factor. In the mass production area of the industry a difference of 1% in fabric utilisation can have a 15-20% effect on the profit made in the making of the appropriate garment. A 3% difference in fabric utilisation will be the difference between a profit making or loss making line.

The process of determining the optimum lay out of the pattern pieces on a piece of material is called 'lay planning' (1). In order to obtain maximum fabric efficiency it is common to put pattern pieces of a number of garment sizes in one lay, and for the lay to consist of the pieces required to make up a number of garments. In order to make the production process more economic, lays are cut in a multi-ply state, i.e lengths of fabrics layed on top of each other. Before the component pieces of a garment, as produced on a layplan can be transformed into an article of clothing by the making up stations the fabric must be selected, spread and cut into the required shape.

Current automated technology allows the cutting of fabric into garment piece parts to be performed an order of magnitude quicker relative to manual techniques. The transition of cut pieces into the sewing room, however, is still labour intensive and requires an inordinate level of handling which influences productivity. This paper describes the current state of progress of a research team within the CIMTEX project. The stripping cell is concerned with the integration of the cutting and sewing rooms, a link that currently does not exist in garment manufacture. Traditionally cut fabric is produced in large volumes prior to make up leading to high levels of work in progress. The old philosophy of laying fabric long and cutting deep is no longer appropriate as the industry is moving towards smaller batch sizes because the number of styles produced in a season is dramatically increasing. The research task of the stripping cell was to automatically unload a cutting table and present the pieces to a conventional transportation system to route the garment piece parts around the sewing room.

3 STRIPPING CELL FEASABILITY AND SIMULATION

As an aid to determining the feasability of automating the stripping process, a robot cell, based on a conceptual design strategy was modelled using the GRASP simulation package (2). The model provided a tool which allowed some quantitative judgements of the prototype system to be made. To put the simulated cell performance into context the current manual system which is in operation throughout industry was analysed.

When manual stripping of the cutting table is performed, garment pieces are not removed from a stack individually, rather they are gathered into bundles of like pieces with the relevant marker from the layplan placed on top. They are then sorted before being passed in bulk to the sewing room. This production philosophy is known as the Progressive Bundle System (PBS) and leads to handling times of between 70-80% in sewing. Thus, for manual stripping relative to the PBS the number of plies is not a significant contribution to the total stripping time. The bundles are usually transported via a skip to the appropriate making-up stations. For automated stripping to compete with manual stripping for a PBS the following condition should hold :-

$$N =< \frac{Ts}{Ta}$$

The initial indication thus being that to justify automatic stripping, the number of plies must be less than or at most equal to the ratio of the time to remove a stack manually to the time to pick a ply automatically. However, in practice this is not the case. Consider the following sequence of events which apply to the manual stripping process but not restricted to the vicinity of the cutter :-

(i) Strip stacks

(ii) Tie and label stacks

(iii) Intermediate storage

(iv) Transportation

(v) Sort stacks into garments

(vi) Seperate plies at the sewing station

Contrast this to the automated stripping process using the unit production philosophy i.e removing all parts constituting one garment from the cutter and load these onto a conventional transportation system to feed sewing stations :-

(i) Find appropriate piece

(ii) Seperate ply

(iii) Load hanger

It can be seen that the automated process replaces a number of labour intensive steps. Any evaluation of the manual process must take these steps into account. Our analysis included two factors, these being :-

(i) The time taken to sort stacks into garments

(ii) The time it takes to seperate the plies for sewing to take place

Transportation time being assumed equal in both cases but ignoring any intermmediate storage.

To enable us to evaluate these factors and obtain a quantitative measure of the time to manually strip, a sample of 50 lays was analysed from the manual production system at CIMTEX. From this sample it was determined :-

(i) The mean strip time per stack was 41.52 seconds from the cutting table

(ii) The mean sort time per stack was approximated at 15 seconds

(iii) The mean ply seperation and load time per garment piece was 15 seconds

Assuming a lay containing four garments, each garment comprising of four pieces and there being twenty plies, the approximate manual stripping time was calculated at ninety five minutes.

The GRASP model was used to provide figures for the automated system. Over a sample of five hundred picks, covering the full range of the stripping table, the average piece pick time was found to be twelve seconds. Thus it can be calculated that for the same lay using the same figures as for the manual system, the automated stripping process would take approximately 64 minutes. These results indicate that automatic stripping of garment pieces can be justified against the subsequent redundant activities of the manual process. This being the case then the actual overall time spent stripping will increase thus reducing the utilisation of the cutter.

This initial analysis of automated stripping has provided positive results. The current trends within the garment industry provide further encouragement to automating the stripping process. Due to its nature the garment industry is rapidly changing and the current trends indicate that batch sizes are reducing with retailers requiring quick response as fashions change. A consequence of smaller batch sizes is less plies per lay. As the number of plies decrease the manual stripping time does not decrease significantly whereas automatic stripping time does. This relationship thus creates a potential bottleneck in the manual system, the elimination of which becomes increasingly significant as UK manufacturers strive to become competitive with their far eastern counterparts.

4 THE ROBOTIC STRIPPING CELL CONCEPT

A prototype stripping cell has been established at CIMTEX which integrates conventional automated CNC cutting tables with traditional Unit Production transportation Systems (UPS) used to route the piece parts constituting a single garment around the sewing room. The cell (Figure 1) consists of two custom built robots operating in a shared workspace. The CAM Systems robot (CSR) is a cartesian device having seven degrees of freedom. It is gantry mounted and operates in a cuboid space (2.5m * 5.0m) above the cutting table. The robot is capable of transporting a payload of 25kg at a speed of 3m/s. The high payload defined is to accomodate the fixed automation necessary to provide the control of the hanger relative to the ply seperation device. The Offstation Loader robot (OLR) is a beam mounted cartesian device with five degrees of freedom.

The basic operation of the stripping cell is that empty hangers arrive via the UPS at the OLR. The OLR picks an empty hanger and moves into the workspace of the CSR. The CSR takes the hanger and picks the required garment pieces before returning the laden hanger to the OLR which passes it back to the UPS where it is routed to the appropriate sewing station.

The CSR end effector and UPS hangers have been designed specifically to facilitate the stripping operation (3) . The hangers, which can hold up to six individual garment pieces are passive and each garment piece is held at two points on the hanger via clamps activated during stripping. The end effector has two functions, namely to perform the pickup of individual components and to control the hanger relative to this process.

5 PLY SEPERATION

Much work has been done on grippers for handling flexible material. These can be broadly categorised as pinch, needle or adhesion techniques. For the purpose of providing a generic, damage free solution where only access to the top of the stack could be achieved an adhesive solution was considered. The research team at Leicester investigated the use of a high tack version of a DTI SMART award winning solid adhesive called MAGNATAC. The innovative aspect of this product was the fact that the adhesive is washable and contamination can simply be wiped off.

It was established that the device must utilise the strength of the adhesive in shear and avoid any peel forces where the adhesive is at its weakest. To this end a roller concept was adopted which aided the ply seperation activity acting to break any fibrous interconnection between plies created during the cutting process.

A series of experiments were set up to determine the significance of contact load, contact time, contact area, seperation rate and the number of seperations. The results of these experiments were used to define theoretical dimensions of the roller device.

An initial prototype production ply seperation device has been constructed. It consists of two short rollers one of which can be moved laterally which enables the device to cope with piece parts of variable width. The rollers themselves are driven by two synchronised stepper motors which prevent the fabric rolling back once seperated as well as preventing distortion of the stack when rolling across it due to the reaction forces associated with undriven rollers. Subsequent testing has led to the requirement of a clamping device to fix the fabric to the roller in order to make the process more robust when dealing with different sized piece parts.

Preliminary testing has indicated that the process is 95% reliable over 500 picks. Although reasonably high this reliablity must be significantly improved before it could be implemented in a commercial environment.

In order to optimise the picking process a number of sensory requirements are necessary. To make use of the pressure sensitive characteristics of the adhesive with use it is necessary to be able to assert a controllable force on the top ply of the stack. Experimentation with strain gauges located on the roller stem has shown these to be adequate for this task. In a production situation the end effector will have to detect whether one ply, no ply or multiple plies have been picked. A simple sensor capable of detecting all three situations is ideally suited. Investigation into the use of reflective, diffuse, optical sensors has shown that trigger levels require constant readjustment with different fabrics. Current work is looking at the suitability of pressure sensors and linear variable differential transformer displacement transducers for this task.

6 STRIPPING CELL CONTROL

The simulated robot cell described earlier only considered the use of one stripping head and paid no attention to optimising the usage of the robots or to the system performance under fault conditions, defined through sensory feedback from the end effector. This section introduces the structure of the stripping cell controller which is being designed to cater for these situations.

The control of the stripping cell is a complex task. The cell is highly interactive with other controllers in the production system. It must also be capable of controlling two robots with a common operating envelope to improve the performance of the stripping function. Figure 2 shows a simplistic representation of the data flow for the stripping cell. As can be seen the robotic controller receives information from a variety of sources. The aim of this is to schedule the system so as to make efficient use of resources to complete tasks in a timely manner.

6.1 STRIPPING CELL HARDWARE

It was decided at an early stage to adopt a modular approach to the design of the controller hardware. This has the benefits of not only allowing future expansion, for example if a second stripping head is added the appropriate control module may be added but also by having a multiprocessor environment the computing effort may be suitably distributed. The I/O requirements of the controller were varied and included DC servo motor drives, stepper motor drives, optical incremental shaft encoder inputs, 24 volt digital outputs, digital inputs and analogue input and outputs.

An IBM 386 PC was selected as the heart of the cell controller. Via an ethernet network the controller could thus interface to all the other cell controllers and computer systems on site. For robot motion control a PCbus based Programmable Multi-Axis Controller (PMAC) was selected. The heart of the PMAC is a Motorola DSP56001 general purpose 56-bit fixed point, 62 instruction digital signal processing chip operationg at 10 MIPS. A single PMAC can control up to 8 axis of motion, either independently or in a synchronised, coordinated mode and provides a range of digital and analogue input/output. Additionally, up to 8 PMAC's may be daisy chained to provide control for up to 64 axis, thus satisfying our criteria for expandability. The function of the PMAC within the cell is shown in Figure 3.

The OLR is a pneumatically operated device which operates in a 'bang-bang' manner. Proximity switches are used to detect each axis position and the drives are switched on/off accordingly. The OLR will always operate in one of two set sequences, either fetch empty hanger from UPS and pass to CSR or fetch loaded hanger from CSR and pass to UPS. It was decided that the simplest and most cost effective way of achieving this sequence control was to use a simple programmable logic controller (PLC). The appropriate sequence would be triggered via the motion control program running on the PMAC. Similarly the end effector stripping process has a requirement for a simple digital I/O sequence. The PLC controlling the OLR was used to provide this, again the sequence being trggered from the PMAC motion control program.

The final I/O requirement for the cell is the end effector sensor information for the monitor module. This being simple analogue and digital input and was achieved via an industry standard PCbus based data acquisition module.

The PMAC has a number of inbuilt safety features. For example, motor following error limits can be configured and if these limits are exceeded the system is brought to a halt in a safe manner. This information may be relayed to the PC host if required. The PMAC has an onboard Intel 8259 programmable interrupt controller integrated circuit. This device has eight inputs that can cause it to send an interrupt signal to the PC. Our system has been configured to generate interrupts on warning following error and fatal following error.

6.2 STRIPPING CELL SOFTWARE

The software comprises of three modules as shown in Figure 4.

(i) Scheduler - Upon a job being entered into the system the stripping controller first determines whether an existing schedule is available. Assuming one is not, information is gathered to produce an appropriate schedule. The major functions of the scheduler are to check for resource availability, assign resources, sequence the jobs and respond to feedback information from the cell. The cut order planning and production control systems provide the required information associated with a particular job (e.g date required by) whilst the UPS provides resource information associated with the overhead conveying system

(ii) Motion Planner - The motion planner assigns actions to a robot, orders the actions, generates collision free trajactories for the robots and synchronises motion between the automatic cutter the robots and the UPS.

The starting point for the motion planner is always a cut file which defines the shape and location of the garment pieces on the cutting table. The cut file which is in ISO format, contains all the necessary instructions for the CNC cutter. The first stage of the motion planner is to filter this ISO file and extract the number of stacks together with the xy coordinate definitions of the edges of each stack.

Having determined the location and shape of each stack the motion planner must determine the pick up point of each garment piece within a stack. The current system allows this to be acheived either automatically, whereby a line along which a piece may be picked is generated by the system, or interactively, in which case the operator is presented with the outline of the stack on the screen and uses the mouse to select the required pick up point.

From the set of generated pick up coordinates, a set of target coordinates for the CSR are calculated. Before a motion program can be generated from the target coordinate definitions, the order in which the garment piece parts are to be picked must be determined. Each hanger holds at most six garment pieces. It is required that a hanger should contain only those pieces pertaining to a single garment. The information relating garment pieces is not contained within the ISO cut file, so at this stage of development must be entered manually.

Robot trajectories may now be generated for the complete stripping sequence. A simplistic approach is taken here to account for the two robots operating within a common workspace. This is necessary because the OLR operates in a 'bang-bang' manner and hence has no real control when moving. A common approach reported in the literature (4,5) is to allow only one robot at a time within the common workspace. However, this approach cannot apply for our application since the hanger must be passed between the CSR and OLR. To overcome this difficulty the following restriction is placed on the OLR. The OLR may not enter or leave the common workspace whilst the CSR is moving. This method is not optimal, but serves to illustrate that automatic stripping is achievable and provides a medium against which alternative algorithms which will form the basis of future research may be evaluated.

The final stage of the motion planner is to generate a motion program for the PMAC. An interpreter has been developed which accepts as input the required robot trajectories in the required order and generates a complete motion program that contains all the instructions necessary to perform an automated clearance of the cutting table. The generated program is passed, under interrupt control, from the PC to PMAC ready for execution.

Once a motion program is executing the motion planner will also respond to information from the monitor. The monitor generates an interrupt if an exception condition is detected. This invokes a rule set within the motion planner in order to establish a correcting course of action.

(iii) Monitor - The monitor provides feedback information relating to the current state of the system. The cell has a number of uncertainties associated with it. Sources of uncertainty in robotic assembly systems have been categorized by Taylor (6) as follows :-

> Components : eg wrong component, component not there
>
> Actuators : eg miscalibrated, joint failures
>
> Fixtures : positioned incorrectly
>
> Assembly process errors : eg more than one ply separated
>
> Sensor errors : eg miscalibrated or not working
>
> Decision errors : e.g if a sensor indicates that a robot failed to pick up a garment piece, how can we tell if the robot is miscalibrated or the piece was positioned incorrectly etc.

The aim of the monitor module is to reduce these uncertainties and provide appropriate information so that errors may be overcome. Information provided by the monitor module includes hanger available on the overhead transport system, faulty material piece, material piece not picked up (from robot end effector). This information is fed back to the motion planner and scheduler modules so as appropriate action can be taken.

7 CONCLUSION

The use of robots for stripping garment pieces is a new and challenging application. This paper has described the current state of progress of a programme of work within CIMTEX and has introduced a methodology for scheduling robots working in real time to perform the stripping operation in garment manufacturing.

Automated stripping is perceived to be a major advance in current clothing manufacture providing the missing link in Computer Integrated Manufacture. The benefits can be seen as :-

(i) reduced handling time leading to increased productivity

(ii) reduced work in progress requiring less capital expenditure on raw materials

(iii) smaller batch sizes due to quick response to demands resulting in a reduced number of garments sold at a discount often constituting significant losses in volume production

(iv) reduced labour intensity and monotonous activity

(v) 100% traceability allowing quality standards to be employed simply

(vi) reduction in factory space for intermediate storage to buffer work passing from cutting to sewing.

REFERENCES

1. Jacobs-Blecha C and Riall W. "The Feasibility of improving The Marker Making Process", International Journal of Clothing Science and Technology, Volume 3 No. 4 1991 pp 13-24

2. Chan S.F. Weston R.H. Case K. "Robot Simulation and Off-Line Programming", Computer-Aided Engineering Journal, August 1988 pp157-162.

3. Paterson A. Kirby N. Hallberg G. "Automated Stripping - A Missing Link In Computer Integrated Manufacture For The Clothing Industry" Presented at Factory 2000, University of York, July 1992

4. Gilbert P.R. Coupez D. Peng Y.M. and Delchambre A. "Scheduling of a Multi-Robot Assembly Cell", Computer-Integrated Manufacturing Systems, Volume 3 No.4 November 1990 pp 236-245

5. Coupez D. Delchambre A. and Gaspart P. "The Scheduling of a Multi-Robot Assembly Cell" Proceedings 5th CIM Europe Conference, 1989 pp185-195

6. Taylor P.M. "Sensing in Advanced Robotic Assembly", Measurement and Control, Volume 23, March 1990 pp 43-46

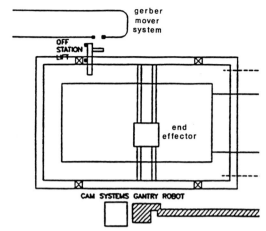

Fig 1   Robotic stripping cell layout

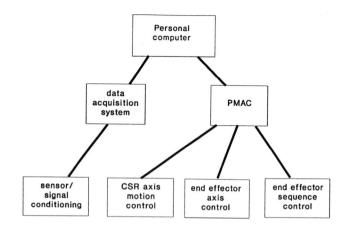

Fig 3   Stripping cell controller hardware structure

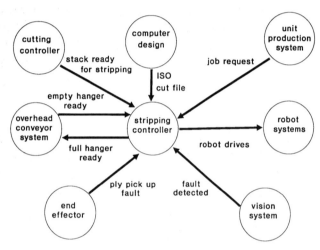

Fig 2   Data flow for robotic stripping cell controller

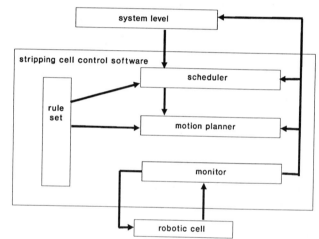

Fig 4   Stripping cell controller software architecture

Sensor integration in high speed packaging machinery

E J RUSHFORTH, BEng, T C WEST, P D KING, BTech, CEng, MIMechE
Loughborough University of Technology, UK

SYNOPSIS

Current production techniques are moving towards in-process measurement, integrated with adaptive control, rather than existing off-line product evaluation. This paper presents the challenges of sensor integration in a new generation of adaptive controlled, high speed packaging machinery. This has been seen to require a Mechatronic approach through the needs of total design integration. The introduction of such sensor based control is seen as the key to a successful new product, providing the packaging manufacturer with perceived needs in terms of dimensional quality, package strength and print quality. The work has provided considerable challenges in designing for adverse environments, high speed materials handling (>10m/s) and flexible raw materials/product.

1. Introduction

Sensor integration within high speed machinery is rather more than just attaching, 'off the shelf' transducers onto an existing machine and hoping to solve all problems. Sensor systems, as such, are defined as transducers with appropriate signal processing, amplification, etc., to provide direct inputs into, for example, a control system in either analogue or digital form.

Current product quality in high speed packaging machinery is variable due to the unpredictable nature of wood products such as paper, cardboard and corrugated board. As the nature of packaging production is geared to high speed and very low unit cost, off-line quality control is limited to sample testing. This testing coupled with a very unpredictable operating environment leads to a low level of quality assurance. This leads to customer dissatisfaction, returned products and ultimately lost business.

Production must move towards measuring the process, rather than the product after manufacture. Research is leading to measurements being carried out on-line within critical production operations such as sheet printing, product handling, and package construction. On-line sensory integration with microprocessors, C.A.D. databases and machine actuators will provide the required adaptive control system.

By successfully carrying out sensor integration, product quality will improve and machine set up time and down time will be reduced substantially. As customers are requesting smaller batches and a higher quality product; design integration in high speed packaging machinery is essential.

This paper outlines the extent to which sensor integration can be taken, and describes the challenges that are being undertaken currently at Loughborough.

A typical high-speed machine for packaging manufacture is shown in Fig 1., manufacturing corrugated board boxes. Fig 2. illustrates the quality problems encountered during such a manufacturing process. Fig 3. is a schematic diagram showing the general processes in packaging manufacture.

2. HIERARCHICAL LEVELS OF SENSOR INTEGRATION

Integration, in this sense, is defined as combining parts to form a whole system, through a multi-disciplinary approach to the design thereof. Sensor integration can be achieved by combining one or more transducers with other devices to form a functioning system. The extent to which integration can be carried-out is constantly increasing with technology. There are six levels of integration to be considered:

2.1 The tangible level

The tangible element being the transducers / sensors physically mounted on or within a machine. The positioning of a sensor must take into account many factors. Apart from the sensor being small enough, it must be capable of withstanding the surrounding environment. Temperature, humidity, dust, vibration, fluid splashes and electromagnetic fields are all potential hazards within packaging machinery. The method of supplying a sensor with power and receiving electronic signals back may be another important consideration especially if the sensor is itself, moving. One major criteria for sensor position is its ability to measure the quantity it was assigned to measure in terms of:

- How accurate will the readings be?
- How repeatable are those readings?
- Does the reading even represent the required quantity?

Often a physical quantity may be impossible to measure and the designer must devise an indirect method.

2.2 The Intangible level

The intangible element being the sensor signal being interpreted into more meaningful data. This is achieved by combining the sensor with a data processing unit. This processing would normally be carried out remotely, however future sensors will have their own localised intelligence (i.e. Smart Sensors [2]). Smart sensors will hopefully facilitate the construction of multi-sensor systems in the future. Currently localised intelligence is generally limited to fixed amplification and conditioning of the signal.

2.3 Integration between sensors

Integration is to provide data which cannot be provided by one stand alone sensor or where it is cheaper to use more than one type of sensor to provide the required information. The sensors may provide similar information (eg two CCD cameras to provide 3D vision) or complementary information (eg. ultrasonic proximity measurements calibrated with air temperature readings to produce more accurate information).

2.4 The control level

The control level is intended to integrate sensor data / information with actuators and control elements, to provide a feedback control system. This level of sensor integration is used widely in Mechatronic machines. A typical, simple, example is that of an optical encoder on a servo motor for accurate positioning.

2.5 On a machine/process level

This level is intended to integrate by controlling and interpreting the signals of many sensors within a machine, or manufacturing cell; using a central controller. This could run decision making algorithms based on competing sensor information. The resulting information will then be sent to actuator control electronics which will initiate actions appropriate to machine status and tasks. This is a real-time activity and requires fast information procurement and processing. There is a minimum amount of human intervention, if any, as all decisions are made by the controlling computer. There can be a capability of self diagnosis of sensor/system failure. When a controller receives data indicating a malfunction or error it can flag an interrupt or alarm signal.

2.6 On production planning and monitoring level

This form of integration could take many forms. The setting up machine parameters and sensor set up, by using data from the planning department (eg. C.A.P. information such as batch quantities, C.A.D. information such a product dimensions). Information provided by the sensors during production can be stored in a database, common to all the machines in the factory. Off-line processing of database information would be used to monitor individual machine performance and overall factory performance. The database can provide statistical process control with aim of continually improving production (J.I.T. philosophy).

3. DETECTION OF BOARD THICKNESS

High speed packaging machinery relies heavily on contact drive rollers to push, cut, crease or print sheets (be it plastic, paper, card or board), in the machine. Each time a production run is changed, adjustments must be made to all roller gaps to ensure the rollers are far enough apart to allow the product though, and close enough to apply the correct pressure to the sheets for optimum print quality and minimum distortion. These adjustments must be made to very fine (10μm) tolerances to ensure quality of print and avoid roll marking on the sheets. This is particularly important in flexographic printing were roller pressure influences print quality significantly.

An off-line thickness sensor can be used to set the roller gaps automatically before each production run. Automatic set up is becoming increasingly important, with the increasing implementation of JIT production. Customers who employ JIT methodology are demanding that their suppliers do the same, by having shorter production runs and faster set up times. The off-line sensor is connected to the central machine control processor.

An on-line thickness sensor caters for variations within each batch which is common when considering quality critical variations of around 10μm. The rollers can only be adjusted slightly during a production run so it is necessary to have both sensors.

For the on-line thickness sensor integration is not so straight forward. On the tangible level, the sensor must be compact, withstand the dusty environment, have a virtually infinite life-span due to the discontinuity of the product, and be capable of operating in a high-speed (>10 units/s), real-time situation. The sensor must be positioned at the front end of the machine to give the control system the largest available leadtime. It is, however, undesirable to have print rollers, continuously adjusting whilst printing, so direct feedback control is unsuitable.

When deciding what sensor or sensors to employ, the main problem is that of speed. A sensor may be able to produce an accurate reading in a fraction of a second, but its frequency response may be too low for it to react quickly enough to a transient change, ie. next production unit.

All contacting sensors have been eliminated from the investigation because their life expectancy is limited to a finite number of operations. For example a device with a life expectancy of 10^7 will last 23 days of continuous production at 18,000 units/hour. Non-contacting sensors fall into four main categories : capacitive/inductive, ultrasonic, pneumatic and optical. The accuracy of capacitive/inductive sensors is effected by material variation. Ultrasonic sensors using air as a the propagation medium are expensive and not widely available at the required accuracy. Pneumatic sensors suffer from a poor response time. Optical triangulation devices which have adequate frequency response are expensive.

New sensors are constantly coming onto the market and one may soon provide the solution. However, new sensors can be prohibitivly expensive. Alternatively a sensor will have to be developed specifically for this task.

Investigation into this problem has highlighted the importance of taking into account all aspects of the problem and the importance of considering the end user's (ie. the machine operator's) requirements.

4. PRINT REGISTRATION AND COLOUR RECOGNITION

As market forces require the retailer to adopt more appealing packaging, there has been a greater demand for higher quality print for point of sale packaging. Consequently finer print stereos (the rubber print impression plate), better quality inks and smaller print registration tolerances are required. Stereo manufacturers and ink manufacturers can now achieve the standards required. However, the packaging manufacturer must currently expend a large amount of time and operator skill to achieve the registration tolerances required.

4.1 Print Registration

During the set up of a packaging machine, one of the must tedious and time consuming tasks is to adjust the "phase" of each printing station in order that every stereo roll is correctly positioned relative to the others and the rest of the machine's parameters (eg feeding of sheets, cutting and creasing etc). This is usually accomplished by:

- manual approximate alignment
- running a few sheets through and assessing error
- adjusting the stereos for any misalignments
- repeating this loop procedure until the registration is correct

This is a slow and wasteful process which must become computer controlled if finer print registration and quality is to be achieved economically.

Automation of print registration set up can only be achieved by sensory integration. Two CCD (Charge Coupled Devices) cameras are positioned at the end of the print section. Each camera is located so that it can see print registration marks as they pass through the machine.

The print registration marks are normally located in the corners or at the sides of sheets, where they will either be removed in the cutting section, or hidden once the package is erected or assembled. The mark consists of cross-hairs which can vary in design and position on the sheets. System automation would be simplified if a universal registration mark is adopted. The most appropriate design would be a cross-hair for each colour stereo, in each sheet corner. Any misalignment, either in terms of phase between each colour or transverse misalignment can be observed from the registration mark.

To ensure total print registration accuracy, there is a camera either side of the sheet to detect small stereo rotations. In order to detect which stereo is at fault, colour detection must be used. Colour detection could be difficult when using a monochrome camera, especially if there are many stereos or if two or more colours are similar (eg black and brown). This problem can be overcome by replacing one or both monochrome cameras with colour cameras. This will, however, increase the system cost.

The main integration challenge will be that of speed. Image processing needs to be quasi real time. Hence, image capturing and image processing algorithms must be fast. In order to capture the moving image of a print registration mark, stroboscopic lighting is required. Parallel processing will be used to carry out high speed image processing of registration marks. A further increase in image processing speed can be achieved by having an intelligent frame-store or a dedicated frame-store for each camera. The later, will allow both images to be processed simultaneously, which will again increase system cost. Once processed, error can be quantified and, with full system integration, machine parameters adjusted.

4.2 Colour Recognition

Retailers are increasingly tightening the standard of colour reproduction for their packaging. Normally during machine set up an operator will take a sample from the first run and compare its colour(s) with colour sample(s) provided by the customer. For a more accurate comparison the operator will carry out the comparison in a 'light box'. This has various types of lighting sources which simulate different types of lighting conditions (ie. daylight, sunlight, warm white fluorescent light). If two colour samples match each other for every illumination then the colour match is correct. Some standards are now so high that the human eye can no longer tell if a colour sample is within specification. As a result packaging manufacturers are having to use either colorimeters or spectrophotometers to quantify colour.

In order to speed up colour measurement an on-line system is to be incorporated into the machine at the end of the printing section. The on-line system will in fact consist of a colour CCD camera located with the print registration cameras. The operator will have access to a knowledge based system (KBS) which contains up to date information on colour readings for all the colours used in the factory. They will choose the colours to be used with each stereo while entering in information for the machine's set up. Readings obtained from the camera will be compared with information obtained from the KBS.

As with print registration, the camera will be made to observe carefully positioned symbols or colour test samples. These consist of every ink colour used. On-line colour recognition can only aid machine set up and act as an alarm system during production to assure quality for the customer. At present it is impractical to integrate this system into the ink mixing process. This is largely due to the complexity of altering ink colour.

It is ultimately intended to integrate the colour camera requirements for both print registration and colour recognition to one camera for each side of the machine.

5. DETECTION OF SHEET SKEWING

If sheets enter a packaging machine are skew, or become skew during travel through the machine, the sheet will have to be scrapped and may possibly cause a machine jam, halting production. The most common cause of skew being due to the unequal distribution of print contact surfaces on each print unit, causing a net sideways force component through the machine. Sheet skew can easily be detected by having two edge detectors located either side of the sheet. An edge detector can be a simple LED/Photodiode system. When the LED beam is blocked by the front edge of the board, the photodiode output state changes. By comparing the time lag between receiving the two output changes, sheet skew can be detected, and the amount and direction of skew can be calculated. By integrating the sensors with the machine's discrete drive units skew, can be corrected effectively. This would, of course, require the machine to have discrete drives to each side of the print stations.

6. CONCLUSION

It can be seen that the aims of sensor integration within high speed machinery are manyfold including:

- Improved product quality
- Increased machine throughput
- Minimum scrap
- Shorter lead times / set up times

The key to success is considered to be that of design integration with a multi-disciplinary 'team' working on the project. This is the very heart of Mechatronics as we see it.

The ultimate aims of this work are that of 'total product control' encompassing production / machine optimisation, statistical process control methods and quality assurance. Sensor integration is part of a total package comprising reliable higher speed machinery and automated set-up.

7. REFERENCES

[1] ROTH,N ,MENGEL,R Sensor Integration - Getting the whole picture. <u>Sensor review</u>, 1992 No.1, Vol 12. pp. 28-33

[2] DAVIES,B Sense & Sensibility: A mater of interpretation. <u>Sensor review</u>, 1992 No.1, Vol 12. pp. 14-16

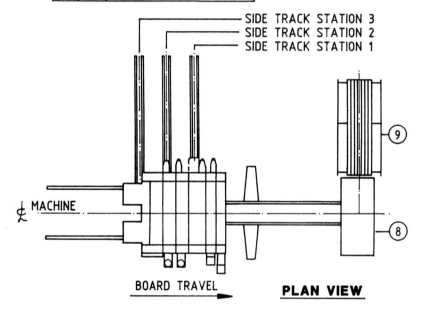
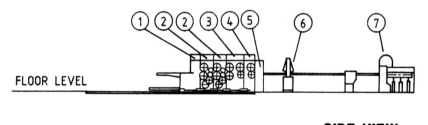

Fig 1 An example of high speed packaging machinery, a corrugated board case maker

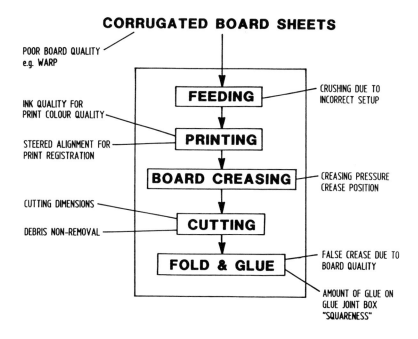

Fig 2 Flow diagram showing the quality problems encountered in corrugated board manufacture

SCHEMATIC DIAGRAM
SHOWING CORRUGATED BOX MANUFACTURE
IN CASE MAKER

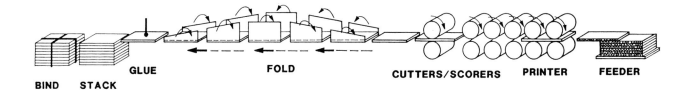

Fig 3 A schematic diagram showing the general processes used in packaging manufacture

Modelling and control of an active actuator system: slow and fast modal effects

SHENGBIAO LI BSc, PhD, **R BARRON,** PhD, ARCST, MIMechE
University of Strathclyde, UK

Abstract:

Active control of vibration has been applied successfully in the areas of rotorcraft and vehicle suspension systems. The integrated nature of these engineering systems places them quite naturally into the subject area – mechatronics.

In active isolation systems, actuators, vibration sensors, and controllers play vital roles. Integration of these devices produces a wide range of dynamic problems. For example, slow and fast modal behaviour has to be considered, because the dynamic characteristics of a complete system can vary over a wide frequency range. The study of engineering systems of this type involves problems of both dynamic analysis and control system design.

To provide an efficient, well designed active system, care has to be exercised in the choice of the dynamic characteristics of the individual components and the specification of a controller algorithm.

Established research work has produced a scheme for selecting these features by the use of modelling and control of both rotorcraft and vehicle suspension systems. In particular slow and fast modal effects have been studied.

Results from simulation studies are shown for a vehicle suspension system. Various design criteria are discussed for a controller and the overall system design characteristics employing singular perturbation and optimal control theory. The results emphasis and show all the important stages and features of the design approach which can be applied to a range of mechatronic systems.

1. Introduction:

In a numer of important integrated engineering systems, for example, an active rotor isolation system and vehicle suspension, there is an increasing need for well documented design procedures. The design of these systems has received considerable attention in recent years because of the need for vibration reduction in order to produce a high system performance and improvements in passenger comfort etc [5, 8]. Techniques to accomplish the design are basically categorised into passive control and active control approaches. Passive control techniques usually involve structural modifications and system parameter optimisation procedures. This technique became more complex and costly in the present stage and the resulting system performance is often limited in comparison with that of an active system. Active control design utilises a feedback control system which incorporates a sensing network for measuring the structural vibrations and a controller which calculates the control input signals for the servo-controlled actuators to reduce the vibration present. An important feature of an active system is that its overall system performance is much superior than that of a passive system and the design techniques of active control have become a dominant research area in recent engineering practice.

In general an active system can be regarded as an integrated mechatronic system with a wide range of dynamic characteristics of active elements. The separation of these system dynamics is very important in order to achieve less interactions in the system and this problem has not been discussed in the recent publications [8]. In particular, the simplifications in system modelling and controller design using the singular perturbation methods also have not been investigated. This paper is intented to present a straightforward solution in relation to the modelling and control of slow and fast modes in an active system. Considerations have been addressed in conjunction to the selection of the dynamic characteristics of active elements. A new criteria of selecting the scaling factor ε has been proposed. Methods for reshaping the frequency response function of closed-loop systems are also investigated and results from the numerical studies verify the approaches discussed in the paper. Furthermore a complete design procedure of an active system has been suggested and is recommended for the design of a wide range of integrated active systems.

2. Slow and Fast Modal Behaviour:

Engineering systems with integrated elements have combinations of structural parts, actuators and sensing devices. This leads to variations in the dynamic characteristics of the elements, in particular to natural or resonant frequencies of vibration. Often producing modes of vibration which are widely separated, having no influence on the dynamic behaviour other than in a known frequency band. Active suspension systems are a good example of this type of problem. Where vehicle/suspension natural frequencies are considerably lower than those of an actuator servovalve or measuring transducer; especially if it is an accelerometer or velocity transducer. With this in mind it is possible to reduce the order of difficulty of the mathematical model, describing the engineering system under consideration, for design proposes. From practical and economic considerations, this is a worthwhile goal and the type of approach used in this work is now described with a view to making it attractive to a wide range of integrated engineering systems.

'Slow' and 'fast' behaviour, in relation to an integrated engineering system depends on a) the physical form of the system and interconnections and b) the dynamic characteristics of individual components and coupling effects in the complete system.

The vehicle suspension system on Figure 1, identifies some of the problems found in integrated systems. In this case, the input disturbance from the road may contain disturbance with a mixture of low, medium and high frequency components, arising from the road conditions. In the direct path of the disturbance the wheel pneumatic tyre system is sensitive to low frequency disturbances (first order effects up to approximately 20 Hz). Next the passive suspension/sprung mass system responds to disturbances in the range from 2 to 5 Hz. By adding the active suspension, the combined actuator/servovalve resonance is in the range up to 10 Hz; the servovalve resonance alone being 2 or 3 times higher.

All electronic amplifiers, microprocessor circuitry and in particular measuring transducers (for example accelerometers) have resonant frequencies in a range from 0.1 KHz to 20 KHz. Thus it is clear that a highly accurate model of the integrated system has to include these effects, if a total design exercise is to be attempted. However since the disturbance inputs of interest are generally in the low to medium frequency range (i.e. < 0.1 KHz), the mathematical model representing the system must accurately cater for the subsystem dynamic effects related to; a) the wheel/tyre system, b) passive and active suspension elements and c) the servovalve and associated electronic circuitry. Where groups a) and b) fall into the 'slow' mode category and c) the 'fast' mode category.

In reducing the analysis of the system to operate in the categories stated, the complexity of the mathematical model may be systematically reduced by using the singular perturbation method [4, 6]. This allows a mathematical model to be produced for design purposes on which simulation studies, to define controller parameters, can be carried out; a procedure to be dealt with in the subsequent sections.

3. Design Considerations For Control of Slow and Fast Modes

Active suspension system design is best illustrated by Figure 2. At the early stage of the design, it is necessary to simulate the system responses of the passive suspension system based on either a simple 2-mass model or a more complicated three-dimensional model. Once the system model is selected, a system optimisation procedure can be carried out, such that the optimum passive suspension parameters are evaluated. The optimisation of the passive system must be carried out since the results of this procedure are used as the comparison with those for an optimal active suspension system in the later stages. Logically, the second stage of design will involve the selection of some active elements which include the sensing devices, dynamic compensators and most importantly the servo-controlled hydraulic actuators. At this stage, the selection of these active system components must take account of the 'slow' and 'fast' modal effects. Effectively, a model reduction approach – the singular perturbation method is used in an active system with multiple modes having separate dynamic characteristics. Depending on the performance requirements of an active suspension system, various types of control system configuration can be chosen for satisfactory system performance, in terms of stability margins, disturbance rejection, system response bandwidth, restricted suspension deflection and road holding capability. These system performances can be guaranteed by the use of recently developed multivariable control techniques, ranging from conventional optimal state feedback control to recently developed robust control approaches and H_∞ methods etc. These modern control techniques can be applied, based on either the concept of composite control or the reduced-order control system design. For a composite control system design, controllers are designed separately based on the requirement specifications of the 'slow' and 'fast' subsystems. whereas the reduced-order control system merely depends on the dynamics of a 'slow' system provided that the 'fast' system is asymptotically stable. One of the main advantages of the singular perturbation methods is that the methods provide the possibility of achieving simplified design not only in the early stage of system modelling but also in the control system synthesis. As a result, the requirement for the number of feedback signals can be reduced in the control system implementation. The effectiveness of the active control system can be finally examined by the simulation studies of the controlled system responses, in terms of system bandwidth, vibration reduction and other system performance requirements. These aspects of the design are illustrated later in the paper.

3.1 General Description of Active Suspension Systems

Active suspensions are characterised by the addition of external power sources, such as compressors or pumps, to achieve superior ride and/or handling performance. They are considered as a means of replacing the spring and dampers of passive systems by actuators as part of or a complete suspension. These actuators act as force generators, according to a control law and operate with various transducers, providing inner loop feedback signals, to their controllers and to track faithfully force demand signals determined by the control law. Generally, these combinations, with a properly designed feedback control system, can supply and absorb given amounts of energy; produce local forces as a function of the variables which are measured at points that may be remote from the position of force generation and can be adapted, as desired, by servo-compensators to give the required energy. These force generators are normally inserted between the vehicle axle and the sprung mass. With proper control the transmitted forces are minimised depending on the excitation conditions. For an active suspension system of this type, the system modes include ultra-fast, fast and slow dynamics which correspond to the responses associated with various transducers, electrohydraulic servo-actuators and the vehicle system respectively. The control system design of such a typical multi-mode active system requires the separation of various system modes such that less interactions can occur.

The dynamics of various transducers, (accelerometers and pressure transducers etc.), can be represented as second-order systems with high natural frequencies. Therefore, since these natural frequencies are well above 500 Hz, their dynamics have little influence on the overall system stability. The servovalve under study is a two-stage mechanical feedback flow control valve with relatively high frequency response (usable up to 170 Hz). Its dynamic characteristics can be described by either a second-order or a first-order transfer function [6, 7, 8].

For the active system considered, it is found that there is one large coefficient $\frac{\beta \rho_0}{V_E}$ and very small parameters, for example the time constant $\frac{2\zeta_{sv}}{\omega_{sv}}$, pressure coefficient K_{PE}. All of these parameters are related to the servovalve and actuator dynamics respectively. The low magnitude of the quantity K_{PE}, indicates that the servovalve is normally operating around the origin of the pressure-flow curves, and near this operating point the valve flow gain is the largest, giving a high gain for the system. Also, the flow-pressure coefficient is at the smallest value, giving a low damping ratio for the system. The presence of these parameters, however, produce a very stiff open-loop matrix which is badly scaled or ill-conditioned and consequently results in a singular perturbed model for the following analyses.

3.2 The Singular Perturbation Method

Normally, difficulties in the numerical analysis occur (e.g. large dimensions and ill-conditioning) from the interaction of slow and fast dynamic modes. Usually fast modes and some of the poorly controllable and observable slow modes are neglected in a simplified model, by employing the aggregation and dominant modes approaches [3]. In the singular perturbation method, these stiffness properties are taken advantage of by decomposing the original ill-conditioned system into slow and fast subsystems, in the two separated time scales [1, 3, 4]. According to the selection of a scaling factor (Appendix Case I and II), the dynamics of an active suspension system can be represented by the following standard singularly perturbed equation:

$$\begin{aligned} \dot{\mathbf{x}} &= \mathbf{A}_{11}\mathbf{x} + \mathbf{A}_{12}\mathbf{z} + \mathbf{B}_1\mathbf{U}_c + \mathbf{\Gamma}\mathbf{w} \\ \varepsilon \dot{\mathbf{z}} &= \mathbf{A}_{21}\mathbf{x} + \mathbf{A}_{22}\mathbf{z} + \mathbf{B}_2\mathbf{U}_c \end{aligned} \quad (1)$$

where submatrices $\mathbf{A}_{11}, \mathbf{A}_{12}, \mathbf{A}_{21}, \mathbf{A}_{22}, \mathbf{B}_1$, and $\mathbf{B}_2$ can be defined according to the state-space equation (10). The above equations (1) are in the standard form, which satisfy Assumption 3.2 of the work of Kokotovic, Khalil and O'Reilly [4]. Thus slow and fast modes of an active system can be described by these equations, with known constant matrices. As ε of equations (1) tends to zero, the resulting expression is the approximation of the original system dynamics. Therefore, the selection of the value of ε is vitally important as it affects the validation of a singularly perturbed system. At the design stage, there is no unique selection of the scaling factor ε because it is usually bounded as $0 \leq \varepsilon < \varepsilon_1$. The selection of singular factor ε is normally based on the ratio of the largest eigenvalue of the slow subsystem to the smallest eigenvalue of the fast subsystem. This feature has been investigated and proved to be conservative by the authors [2, 5, 6]. A better way of selecting ε can be achieved if only the time constants of the fast system are related to the selection (Appendix Case I and II) [5, 6]. In this way, the numerical values of each element in the open-loop system matrix of equation (10) are well-balanced. Another advantage of this selection method is that it is unnecessary to calculate the eigenvalues for both the slow and fast subsystems before the determination of ε.

The standard singularly perturbed equations (1) can be decomposed into the following completely decoupled slow and fast subsystems based on the similarity transformations [4].

Slow Subsystem:

$$\begin{aligned} \dot{\mathbf{x}}_s &= \mathbf{A}_0 \mathbf{x}_s + \mathbf{B}_0 \mathbf{u}_s + \mathbf{\Gamma} \mathbf{w} \\ \mathbf{z}_s &= -\mathbf{A}_{22}^{-1}(\mathbf{A}_{21}\mathbf{x}_s + \mathbf{B}_2 \mathbf{u}_s) \end{aligned} \qquad (2)$$

where

$$\begin{aligned} \mathbf{A}_0 &= \mathbf{A}_{11} - \mathbf{A}_{12}\mathbf{A}_{22}\mathbf{A}_{21} \\ \mathbf{B}_0 &= \mathbf{B}_1 - \mathbf{A}_{12}\mathbf{A}_{22}^{-1}\mathbf{B}_2 \end{aligned} \qquad (3)$$

Fast Subsystem:

$$\varepsilon \dot{\mathbf{z}}_f = \mathbf{A}_{22}\mathbf{z}_f + \mathbf{B}_2 \mathbf{u}_f \qquad (4)$$

3.3 Control System Design Considerations:

After the decomposition of the original system dynamics, state-feedback control and other recently developed robust control techniques design can then proceed for each lower-order subsystem in different time scale, and the results combined to yield a composite state-feedback control for the original active system. Meanwhile, a composite controller is required to achieve an asymptotic approximation to the closed-loop system performance which would be obtained had a state-feedback controller been designed, without the application of the singular perturbation approaches.

The composite control for the original system is shown in **Figure 3** and governed by the following relationship:

$$\mathbf{u}_c(t) = \mathbf{u}_s(t) + \mathbf{u}_f(t) \qquad (5)$$

where the parameters $\mathbf{u}_s$ and $\mathbf{u}_f$ can be determined based on the slow and fast subsystem performance requirements respectively [1, 3, 4]. This composite control has two important features: 1) a reduction in computational requirements is achieved by solving two lower-order control problems in separate time scales; and 2) the resulting composite feedback control does not require knowledge of the singular perturbation parameter ε, which may represent small uncertain parameters. Since the control of the fast system occurs in an $\frac{1}{\varepsilon}$ faster time scale than that of the slow system, the response of the original system (1) will be mainly dominated by the reduced slow system state after the decay of fast system transient. Therefore, if the fast system is asymptotically stable, then it is only necessary to stabilise the reduced-order slow system and consequently a reduction in control system analysis and design is achieved. This resulting design of the reduced-order controller is efficient in terms of the number of variables considered and measurements required at the stage of the practical implementation of the control laws.

In practice, the eletrohydraulic isolation system considered, employs acceleration of the isolated mass $\ddot{x}_s$ and relative displacement between the sprung mass and unsprung mass Δx as the primary feedback signals. These signal are modified and combined in the servoamplifier to produce a command signal (that operates a servovalve) to control the fluid flow to and from the hydraulic actuator. The assumption is that the fluid is almost incompressible. In the cases where it is desirable to obtain both resonant vibration control and a high degree of isolation characteristics over a broadband frequency range, electric networks which employ notch and/or bandpass filters for shaping the frequency-response characteristics of electrohydraulic isolation system can be incorporated in the acceleration feedback loop; such that single-frequency and multi-frequency vibration isolation can be achieved. The networks can be represented by a transfer function in the terms of the sprung mass acceleration $\ddot{x}_s$ and network response as:

$$\frac{X_{ci}(s)}{\ddot{x}_s(s)} = \frac{2\zeta\omega_i}{s^2 + 2\zeta\omega_i s + \omega_i^2} \qquad (6)$$

where $i = 1, 2, ...$, $\omega_1 = 2\pi f_1$, and $\omega_2 = 2\pi f_2, ...$. For demonstration purposes, f_1 and f_2 are the specified frequencies, chosen for example as 2 Hz and 5 Hz respectively.

Another important feature, associated with the design of a low-frequency isolation system is that extremely large static and steady-state relative deflections occur due to sustained acceleration conditions for low values of resonant frequency. Therefore, an integral feedback loop of the relative displacement must be introduced if the static and steady-state deflections are to be eliminated. The controller used is a typical state feedback controller; experimental studies on this type of controller have verified that it works in practice e.g. in [8] on a helicopter rotor isolation system. An aim of this work is the implementation of a reduced order state feedback controller which does not require the measurement of the spool displacement and velocity of the servovalve. Thus a simplified controller structure is produced which deals efficiently with active suspension systems and indeed will be less costly to produce than a full state feedback controller. This reduced-order controller structure is schematically illustrated in Figure 1 where the acceleration, relative displacement and pressure signals are employed in the configuration. The controller is basically a microprocessor based controller which includes both A/D and D/A converters for data acquisition and control. The analogue signals from the measurements are converted into digital signals for the control algorithms, to produce the control inputs which are later converted into analogue signals for the suspension system. In a laboratory situation, these A/D and D/A devices can be effectively replaced by a complete data acquisition system (e.g. DAP 1200 from Microstar Laboratories Inc) which occupies one expansion slot in a personal computer. This type of device combines analogue data acquisition hardware with a 16-bit microprocessor, a large buffer memory, and a real-time multitasking operation system. During the data acquisition process it is important that a suitable sampling rate is chosen. However in the field this equipment can be replaced by a dedicated microproxessor system.

Based on the augmented equation (15), the LQR controller design method can be applied for reshaping the frequency response of the closed-loop system. In this case, the performance index is chosen as:

$$J = E[\gamma_1 \ddot{x}_s^2 + \gamma_2 \Delta x^2 + \gamma_3 x_c^2 + \gamma_4 u^2] \quad (7)$$

and in terms of standard integral matrices, the above performance index can be written as:

$$J = \lim_{t \to \infty} \int_0^t e^{2\alpha t}(\hat{\mathbf{x}}^T \mathbf{Q} \hat{\mathbf{x}} + \hat{\mathbf{u}}^T \mathbf{R} \hat{\mathbf{u}}) dt \quad (8)$$

with the matrices $\mathbf{Q}$ and $\mathbf{R}$ defined as:

$$\mathbf{Q} = \gamma_1 \mathbf{Q}_1 + \gamma_2 \mathbf{Q}_2 + \gamma_3 \mathbf{Q}_3, \quad \mathbf{R} = \gamma_4. \quad (9)$$

where the matrices $\mathbf{Q}_1$, $\mathbf{Q}_2$, $\mathbf{Q}_3$ are the weighting matrices associated with the responses of the sprung mass acceleration $\ddot{x}_s$, relative suspension deflection Δx and the bandpass filter state $\mathbf{x}_c$ respectively.

From previous studies [6], it is possible to shape the frequency response of the closed-loop system by selecting different weighting factors γ_1, γ_2, γ_3 and γ_4. Alternatively, a combined LQR and pole placement method can also be applied for this active isolation system such that a desirable frequency response shape of the closed-loop system is achieved. Basically, this combined LQR and pole placement method is conducted in two steps:

(1). The LQR method is applied first to compute the feedback gain matrix $\mathbf{K}$ based on the optimal control problem for a system with an appropriate performance index. Then the closed-loop matrix $\mathbf{A}_c = \hat{\mathbf{A}} + \hat{\mathbf{B}}\mathbf{K}$ and its eigenvalues are computed.

(2). In some cases, the resulting closed-loop eigenvalues from the LQR method are not placed in the desirable positions in the complex plane. Therefore, the frequency response of the closed-loop system is not suitable in terms of vibration reduction, bandwidth of the resulting system and other system performance. This difficulty can be resolved subsequently by the pole placement approach where only the dominant poles need to be repositioned in the complex plane, such that the resulting system performances are satisfactory. In these occasions, the feedback gain vector $\mathbf{K}$ will be recalculated such that only the dominant closed-loop poles are in the prespecified positions. It can also be found that the frequency response of the closed-loop system are closely related to the positions of the dominant poles. Thus, the frequency response of the system can reshaped as desired by altering the positions of the dominant closed-loop poles.

The advantages of this method are that: first of all it is very difficult to make a proper judgement for the desired positions of dominant closed-loop poles and consequently the resulting shape of the frequency response, before the application of the pole placement approach. By applying the LQR method, the resulting closed-loop eigenvalues are evaluated from the solution of the Riccati equations and the closed-loop frequency response calculated; secondly, since only the positions of the closed-loop dominant poles affect the system performance, the feedback gain matrix $\mathbf{K}$ is not changed significantly from its original calculation by the LQR method, if only the positions of the dominant poles are changed. Therefore, it is understandable that the performance of the resulting closed-loop system is suboptimal, since only the dominant poles are repositioned while the rest of the closed-loop eigenvalues of the closed-loop system virtually remain unchanged. Finally, the most significant point of the method is that it provides a systematic approach to the reshaping of the frequency response of closed-loop system, as desired.

4. Numerical Simulations And Discussions:

In the simulations, the weighting matrices $\mathbf{Q}$ and $\mathbf{R}$ are first chosen as follows:

$$\mathbf{Q} = diag(0, 0, 0, 0, 0, 700, 700, 100, 100), \quad \mathbf{R} = 100$$

By selecting different weighting factors γ_1, γ_2, γ_3 and γ_4 in the performance index J, the frequency response $|\frac{F_t(j\omega)}{F_d(j\omega)}|$ of the closed-loop system can be reshaped such that the system performances, in the terms of vibration reduction, system bandwidth and other requirements, are satisfactory. As illustrated in Figure 4, for example, the frequency response of the system is closely related to the weighting factors mentioned. The presence of the bandpass filters provide a high level of vibration isolation at the prespecified frequencies of 2 Hz and 5 Hz respectively. As γ_2 changes, the frequency response $|\frac{F_t(j\omega)}{F_d(j\omega)}|$ change dramatically and the highest rates of isolation of the system are approximately $-37dB$ and $-45dB$ at the notched frequencies with $\gamma_2 = 5$. The relationships between the frequency response function $|\frac{F_t(j\omega)}{F_d(j\omega)}|$ and other weighting factors can be found in [6].

Also with the combined LQR and pole placement method discussed in section 3.3, it is found that the direct solution of the LQR approach gives the closed-loop eigenvalues as: $-170.1 \pm 2611.2j$, $-753.3 \pm 765.5j$, $-1.6 \pm 31.3j$, -5.7 and $-0.7 \pm 12.5j$. In comparison the subsequent simulations for bandpass filters with $\zeta_b = 0.01$ and $\zeta_b = 0.04$, the dominant closed-loop eigenvalues are modified to $-0.7 \pm 0.9j$, $-0.3 \pm 0.5j$ and $-1.2 \pm 1.3j$. In each case, the frequency response function $|\frac{F_t(j\omega)}{F_d(j\omega)}|$ and its phase are shown in Figures 5 and 6 for selected conditions. It is found that the frequency response of $|\frac{F_t(j\omega)}{F_d(j\omega)}|$ has been reshaped dramatically compared with those of the direct solution of the LQR method and the passive system. A force transmissibility less than -14 dB is provided over the frequency range of approximately 1 to 20 Hz. In other words, the vibration isolation performance of the system has been improved significantly for the reduction of vibration with the frequencies ranging from 0.5 Hz to 20 Hz as shown in Figure 5. Finally, the time-domain response of the system shown in Figure 7 indicates the effectiveness of the active system performance in relation to the reduction of vibration when the system is properly controlled.

5. Comments and Conclusions

Both the analytical and simulation work are presented, where the singular perturbation method is applied to the modelling and control of an active vehicle suspension system. Slow and fast modal effects in relation to the selection of the dynamic characteristics of the active elements in this integrated system have been investigated. A complete design procedure for the control of slow and fast modes in active vehicle suspension system is outlined and recommended as a standard approach in mechatronic systems. Simplifications in system modelling and controller design are achieved by using the singular perturbation approach. Validation tests of the reduced-order system model are carried out in conjunction with the selection of the scaling factor ε. A method for selecting the scaling factor ε is shown and recommended for use in integrated system design.

Results (Figures 4 and 5) indicate that reduced-order control system synthesis is viable provided that the fast system is asymptotically stable. Two design methods for a satisfactory controller are applied systematically for a vehicle suspension system. From this approach the desired frequency response for an integrated system can be achieved. A high level of vibration isolation has been achieved by the present design methods (Figure 6). The simulation studies verify the validation of the reduced-order model and associated controller design approaches.

It is recognised that during the motions of a vehicle the interactions between the active suspensions will have an effect on the overall response. For example the heave and pitch motions will be strongly coupled and as such the controller has to be designed to cope with the combined motions. Thus at the design stage a controller has to be designed for each motion separately i.e for the decoupled system. Then the behaviour of each controller has to be refined, to deal with the predominant motion. At the present time the design approach has only been validated for uncoupled or weakly coupled systems. Also the design approach should be applied, before the practical implementation stage of an active suspension system. The improvements in the design of a product can be carried out in this way efficiently and economically and may be supplemented by experimental tests.

Furthermore the methods and results of this work can be applied to a wide range of integrated active isolation systems; and are recommended for use in mechatronic system design.

References

[1] Chow J.H. and Kokotovic P.V., 1976, "A Decomposition of near Optimum Regulators for Systems with Slow and Fast Modes." *IEEE Transaction on Automatic Control*, AC-21(5):701-706.

[2] Feng W., 1988, "Characterization and Computation for Bound ε^* in Linear Time-Invariant Singularly Perturbed System." *Systems & Control Letters* 11, (1988), pp. 195-202.

[3] Kokotovic P.V. and Haddad A.H., 1975, "Controllability and Time-Optimal Control With Slow and Fast Modes." *IEEE Transaction on Automatic Control*, AC-20(1): 111-113.

[4] Kokotovic P.V., Khalil H.K. and O'Reilly J.,1986, "Singular Perturbation Methods in Control: Analysis and Design." *Academic Press*.

[5] Li S. and Barron R., 1991, "Active Vibration Control For a Rotor Isolation System Using a Singular Perturbation Approach." *International Journal of Control and Computers*, Vol. 19, No. 2, pp. 37-44.

[6] Li S., "Control Strategies And Design To Improve The Performance of Active Vibration Isolation Sys-

tems", PhD Dissertation, University of Strathclyde, May, 1992.

[7] Merritt H.E., "Hydraulic Control Systems." *John Wiley & Sons, Inc.*, 1967.

[8] Strehlow H., Mehlhose R., and Obermayer M., 1977, "Active Helicopter Rotor Isolation With Application of Multivariable Feedback Control." *Third European Rotorcraft and Powered Lift Forum,* Aixen-Provence, Paper No.23, pp.1-30.

Appendix

Case I: Variable P_E is a Fast Variable

By scaling that
$$\varepsilon = \frac{1}{\frac{2\zeta_{sv}}{\omega_{sv}}}\sqrt{\frac{V_E}{\beta\rho_0}}$$

and defining the following state variables as:

$$x_1 = \Delta x \ , \ x_2 = \dot{\Delta}x \ , \ z_1 = x_{sv} \quad z_2 = \dot{x}_{sv} \quad z_3 = \varepsilon P_E$$

the singularly perturbed form for the rotor isolation system with the second-order servovalve model can be written as:

$$\begin{bmatrix} \dot{x}_1 \\ \dot{x}_2 \\ \varepsilon \dot{z}_1 \\ \varepsilon \dot{z}_2 \\ \varepsilon \dot{z}_3 \end{bmatrix} = \begin{bmatrix} 0 & 1 & 0 & 0 & 0 \\ -MK_1 & -MC_1 & 0 & 0 & MA_E \\ 0 & 0 & 0 & \varepsilon & 0 \\ 0 & 0 & -\omega_{sv}^2\varepsilon & -2\zeta_{sv}\omega_{sv}^2\varepsilon & 0 \\ 0 & -\frac{\beta\rho_0}{V_E}A_E\varepsilon^2 & \frac{\beta\rho_0}{V_E}K_q\varepsilon^2 & 0 & -\frac{\beta\rho_0}{V_E}K_{PE}\varepsilon \end{bmatrix}$$

$$\begin{bmatrix} x_1 \\ x_2 \\ z_1 \\ z_2 \\ z_3 \end{bmatrix} + \begin{bmatrix} 0 \\ 0 \\ 0 \\ \varepsilon\omega_{sv}^2 K_i \\ 0 \end{bmatrix} U_c + \begin{bmatrix} 0 \\ \frac{1}{M_u} \\ 0 \\ 0 \\ 0 \end{bmatrix} F_d \qquad (10)$$

which is the standard singularly perturbed form of equation (1) with submatrices $\mathbf{A}_{11}, \mathbf{A}_{12}, \mathbf{A}_{21}, \mathbf{A}_{22}, \mathbf{B}_1$, and $\mathbf{B}_2$ defined as:

$$\mathbf{A}_{11} = \begin{bmatrix} 0 & 1 \\ -MK_1 & -MC_1 \end{bmatrix}, \quad \mathbf{A}_{12} = \begin{bmatrix} 0 & 0 & 0 \\ 0 & 0 & \frac{MA_E}{\varepsilon} \end{bmatrix}$$

$$\mathbf{A}_{21} = \begin{bmatrix} 0 & 0 \\ 0 & 0 \\ 0 & -\frac{\beta\rho_0}{V_E}A_E\varepsilon^2 \end{bmatrix}, \mathbf{B}_1 = \begin{bmatrix} 0 \\ 0 \end{bmatrix}, \quad \mathbf{B}_2 = \begin{bmatrix} 0 \\ \varepsilon\omega_{sv}^2 K_i \\ 0 \end{bmatrix}$$

$$\mathbf{A}_{22} = \begin{bmatrix} 0 & \varepsilon & 0 \\ -\omega_{sv}^2\varepsilon & -2\zeta_{sv}\omega_{sv}\varepsilon & 0 \\ \frac{\beta\rho_0}{V_E}K_q\varepsilon^2 & 0 & -\frac{\beta\rho_0}{V_E}K_{PE}\varepsilon \end{bmatrix}$$

In this case, state variables $x_{sv}, \dot{x}_{sv}$, and P_E are defined as the fast variables.

Case II: Variable P_E is a Slow Variable

By the same scaling factor ε but defining the following state variables:

$$x_1 = \Delta x \ , \ x_2 = \dot{\Delta}x \ , \ x_3 = \varepsilon P_E \quad z_1 = x_{sv} \quad z_2 = \dot{x}_{sv}$$

$$\mathbf{A}_{12} = \begin{bmatrix} 0 & 0 \\ 0 & 0 \\ \frac{\beta\rho_0}{V_E}K_q\varepsilon & 0 \end{bmatrix}, \quad \mathbf{A}_{22} = \begin{bmatrix} 0 & \varepsilon \\ -\omega_{sv}^2\varepsilon & -2\zeta_{sv}\omega_{sv}\varepsilon \end{bmatrix}$$

$$\mathbf{A}_{21} = \begin{bmatrix} 0 & 0 & 0 \\ 0 & 0 & 0 \end{bmatrix}, \quad \mathbf{B}_1 = \begin{bmatrix} 0 \\ 0 \\ 0 \end{bmatrix}, \quad \mathbf{B}_2 = \begin{bmatrix} 0 \\ \omega_{sv}^2 K_i\varepsilon \end{bmatrix}$$

in this case, variable εP_E is considered as a slow variable and only variables x_{sv} and $\dot{x}_{sv}$ are treated as the fast variables.

Here, to support the selection of the scaling factor ε, it is found that the determination of the bound ε_1 can be found from the condition; if none of the eigenvalues of matrix $\mathbf{Q}$ of the equation:

$$\begin{aligned}\mathbf{Q} &= \lim_{t\to\infty}(\mathbf{A}_{22}^{-2}\mathbf{A}_{21}\mathbf{A}_0^{-1}(\mathbf{I}-j\omega\mathbf{A}_0^{-1})^{-1}\mathbf{A}_{12}+\mathbf{A}_{22}^{-1})\\ &= \mathbf{A}_{22}\mathbf{A}_{21}\mathbf{A}_0^{-1}\mathbf{A}_{12}+\mathbf{A}_{22}^{-1}\end{aligned} \qquad (11)$$

has a real positive part, then the Nyquist plot of $\lambda(\mathbf{G}(j\omega))$ [2, 6] will not intersect the negative real axis. Thus, the upper-bound ε_1 is infinite. This condition is applied to the system considered and therefore the modelling of the system by singular perturbation method is valid for a wide range of ε.

In control system analysis, the transfer function of the system (6) together with integral feedback of the relative deflection Δx can be rewritten in terms of slow and fast variables as:

$$\begin{bmatrix} \dot{x}_i \\ \dot{x}_{c01} \\ \ddot{x}_{c01} \\ \dot{x}_{c02} \\ \ddot{x}_{c02} \end{bmatrix} = \begin{bmatrix} 1 & 0 & 0 & 0 & 0 \\ \mu_1\frac{C_1}{M_s} & 0 & 0 & 0 & 0 \\ \mu_1\frac{K_1}{M_s} & 0 & 0 & 0 & -\mu_1\frac{A_E}{\varepsilon M_s} \\ \mu_2\frac{C_1}{M_s} & 0 & 0 & 0 & 0 \\ \mu_2\frac{K_1}{M_s} & 0 & 0 & 0 & -\mu_2\frac{A_E}{\varepsilon M_s} \end{bmatrix} \begin{bmatrix} x_1 \\ x_2 \\ z_1 \\ z_2 \\ z_3 \end{bmatrix}$$

$$+ \begin{bmatrix} 0 & 1 & 0 & 0 \\ -\omega_1^2 & -\mu_1 & 0 & 0 \\ 0 & 0 & 0 & 1 \\ 0 & 0 & -\omega_2^2 & -\mu_2 \end{bmatrix}\begin{bmatrix} x_{c1} \\ \dot{x}_{c1} \\ x_{c2} \\ \dot{x}_{c2} \end{bmatrix} \qquad (12)$$

where the quantities $\mu_1 = 2\zeta\omega_1$ and $\mu_2 = 2\zeta\omega_2$. and in a compact form:

$$\dot{\mathbf{x}}_c = \mathbf{A}_{c1}\mathbf{x} + \mathbf{A}_{c2}\mathbf{z} + \mathbf{F}_c\mathbf{x}_c \qquad (13)$$

Combining the equation (1) with (13), the following 10th order equation will be the augmented state-space representation of the closed-loop active vibration isolation system.

$$\begin{bmatrix} \dot{\mathbf{x}} \\ \dot{\mathbf{z}} \\ \dot{\mathbf{x}}_c \end{bmatrix} = \begin{bmatrix} \mathbf{A}_{11} & \mathbf{A}_{12} & 0 \\ \frac{\mathbf{A}_{21}}{\varepsilon} & \frac{\mathbf{A}_{22}}{\varepsilon} & 0 \\ \mathbf{A}_{c1} & \mathbf{A}_{c2} & \mathbf{F}_c \end{bmatrix}\begin{bmatrix} \mathbf{x} \\ \mathbf{z} \\ \mathbf{x}_c \end{bmatrix} + \begin{bmatrix} \mathbf{B}_1 \\ \frac{\mathbf{B}_2}{\varepsilon} \\ 0 \end{bmatrix}\mathbf{u} + \begin{bmatrix} \mathbf{\Gamma} \\ 0 \\ 0 \end{bmatrix}\mathbf{w}$$

$$(14)$$

or:

$$\dot{\hat{\mathbf{x}}} = \hat{\mathbf{A}}\hat{\mathbf{x}} + \hat{\mathbf{B}}\hat{\mathbf{u}} + \hat{\mathbf{\Gamma}}\mathbf{w} \qquad (15)$$

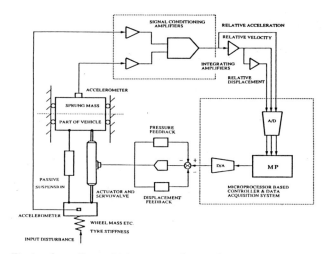

Fig 1 An active vehicle suspension system

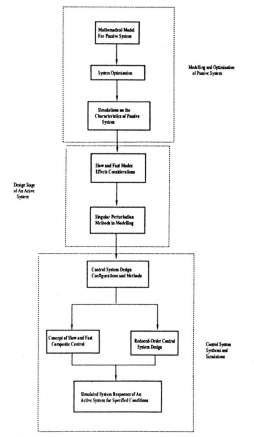

Fig 2 Design procedures for an active isolation system

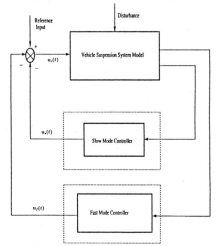

Fig 3 Slow and fast mode control system design approaches

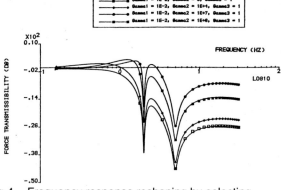

Fig 4 Frequency response reshaping by selecting different weighting factors

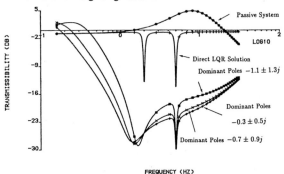

Fig 5 Frequency response by combined LQR and pole placement methods

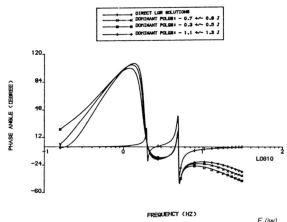

Fig 6 Phase angle of frequency response function $\frac{F_t(jw)}{F_d(jw)}$ by combined LQR and pole placement methods

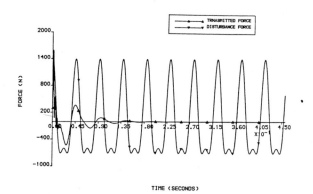

Fig 7 Time-domain response of the closed-loop system

The design of mechanical amplifiers using piezoelectric multilayer devices for use as fast actuators

J K THORNLEY, BSc, T G KING, BSc, MDesRCA, PhD, M E PRESTON, PhD,
Loughborough University of Technology, UK

SYNOPSIS.

The design process leading to the development of a family of mechanical amplifiers which use piezoelectric multilayer actuators as prime movers is described. This includes consideration of choice of materials and the use of finite element and analytical design techniques to arrive at and optimise the design, and ascertain its performance.

It is clear that in the field of actuators, the components available to engineers involved in a Mechatronic design process are mostly of the electromagnetic type, such as solenoids and many variants of the electric motor. Efficient force and displacement transformation is seen as the key to unlocking the potential of applications for modern piezoelectric devices in the Mechatronic sphere. Piezoelectric multilayer actuators can rapidly generate very large stall forces. A typical device measuring 18 mm long by 2 x 3 mm section, will produce a stall force of 200 N in less than 0.2 ms. However, the small, free output movement obtained, typically 15 μm, is almost unusable in medium or low precision machines. Efficient amplification or transformation of such movement, with optimisation of the force-displacement product (at the output), makes it possible to include piezoelectric multilayer actuators into lower precision, lower cost, high-speed machines.

A typical mechanism is described in this paper which generates a 20 Newton stall force, with an unrestrained movement of 100 μm, derived from a piezoelectric device which only produces 15 μm displacement. The device was manufactured and its performance compared to predictions made by a 'designer' program, and is just one in a series of monolithically constructed designs. Although it can not be considered as highly efficient (69%), it transforms the output movement up to a level which can be utilised for applications such as clutching, gripping and latching.

1 INTRODUCTION.

Since piezoelectric multilayer actuators develop maximum displacements in accordance with the electrically producible maximum strain, (typically within the range 600 p.p.m. to 850 p.p.m.), displacement amplification of some type is often advantageous.

The simplest solution to this problem is the obvious one of using a lever system. However, closer inspection of the problem reveals hidden pitfalls associated with hysteresis and wear in bearing surfaces; factors which must be considered when contemplating using pivots with lever systems. The small displacements, typically 10 - 20 μm, and the large potential forces, typically around 100 kgf, eliminate bearing type pivots. However, flexural hinges do not suffer from these problems and can be adopted by the designer in order to facilitate the rotations necessary in lever systems.

2 THE GENERIC SOLUTION.

Figure 1 shows a generalised representation of a flexure hinged displacement amplifier which indicates the topology of the structure, and the variables for which solutions need to be found. The diagram shows linearly tapered beams, and their use is possible, but some justification can be made for the beneficial effect of using beams of elliptical shape. Although this argument is not discussed here, the program was configured to assume elliptical beams.

The design process is as follows;

1) Determine the required output force / displacement characteristics.

2) Choose an actuator.

3) Select a suitable material and billet thickness.

4) Solve geometry for;

   4.1) l_1, l_2, w_1, w_2
   4.2) a_1, a_2
   4.3) d, e_1, e_2

The aim of the design process is to achieve;

i) A high work efficiency, i.e. the ratio of the force displacement product of the output, to that of the input by the prime mover.

ii) Low stressing in all operational modes to maximise fatigue life.

iii) Minimum size and hence mass and therefore maximum speed of response.

The design procedures outlined in this text have been implemented as a 'beam designer' CAD program. This program assists the designer to find an efficient solution based on the following parameters;

i) The material's elastic modulus, billet thickness, yield stress and stress safety factor.

ii) The input parameters of compliance, displacement.

iii) The output displacement required.

iv) Tuning factors associated with compliance matching between various structural zones, selected by the designer, which affect the force displacement efficiency and actuation speed of the design.

3 DESIGN METHODOLOGY.

3.1 OUTPUT FORCE / DISPLACEMENT CHARACTERISTICS.

The requirement of output characteristics is entirely application specific, however it is possible to define a loose working envelope. Firstly, the generic topology can not exist for gains of less than x1. Gains greater than x2 are favoured, with a practical upper limit of approximately x20. In combination with the selection of an actuator, the output displacement determines the gain factor.

The envelope for output force is intimately linked with displacement. A useful criterion for such structures is the force displacement product (See para. 3.2).

3.2 CHOICE OF ACTUATOR.

It is shown later that overall device efficiencies between 50% and 85% are possible, with practical values between 60% and 70%. These figures can be used to select a suitable actuator, also based on the required output characteristics. If the minimum required output force has been determined, in conjunction with the output displacement, this can be referred back to the actuator through the efficiency value. It is conservative initially to select an efficiency of 55% to 60%, and base the actuator selection on this value. For efficiency η, the relationship between input and output forces and displacements is;

$$\eta \, F_{act} x_{act} = F_{out} x_{out} \qquad (1)$$

3.3 CHOICE OF MATERIAL AND BILLET THICKNESS.

A determining factor for the thickness of material is that of actuator geometry. Billet thicknesses less than the actuator thickness are possible but not recommended. Efficient structures favour thick billets with thin hinges, but extremes here are usually impractical. For reasons associated with stress loading and actuator/billet elastic modulus, values of billet thickness of approximately 120% of the actuator thickness are practical.

3.4 SOLUTION OF GEOMETRY.

The structure is considered in two modes of deformation, specifically;

1) The structure is free at the output, with the input fully driven.

2) The output is stalled or vertically constrained, with the input fully driven.

In both modes, the input is driven through an external compliance. This compliance will usually be associated with the compliance of the actuator.

The finite element analysis of structures of this type has shown that maximum stressing always occurs in the flexure hinges, as expected. For this reason, the 'beam designer' does not check for stresses in the elliptical beam members.

3.4.1 SOLUTION OF l_1, l_2, w_1, w_2.

Figure 3 indicates the technique employed to find the widths and lengths of the flexure hinges. The vertical compliances of the main beam are considered to be zero. This is a good working approximation and facilitates an effective isolation of l_1, l_2, w_1, w_2. In mode 2, the response of the structure will be approximately manifested as direct stress in both hinges. These stresses can only be determined by knowing the input compliance of the structure in its currently modelled state. The compliance of the drive is given by;

$$S_d = \frac{x_{i_{free}}}{f_{i_{stall}}} \qquad (2)$$

The input compliance in mode 2 will be given by;

$$S_i = \frac{\left(\dfrac{l_1}{w_1} + \dfrac{l_2}{w_2}\right)}{Eb} \qquad (3)$$

For coupling maximum energy into the structure, the perfect situation would be for the input compliance to be zero. This is impossible, so a *tuning factor* is introduced;

$$k_{stall} = \frac{S_i}{S_d} \qquad (4)$$

which should be minimised as far as possible, to achieve an acceptable overall efficiency, without the geometry becoming unwieldy. This factor is set in the 'beam designer' program. Typical values lie in the range **0.10** to **0.25**. Should the resulting design be too inefficient, the design process can be repeated.

It can be seen that the aspect ratio of the hinges is deterministic for this relation, in;

$$r_{aspect} = \frac{s_i}{Eb} \quad (5)$$

and this can be chosen to satisfy the stall-stress relation. To solve this, the stall force must be determined by;

$$f_{stall} = \frac{x_{i_{free}}}{(s_i + s_d)} \quad (6)$$

Assuming that the hinges have equal width, and are of equal length, then the widths are fixed by;

$$w_1 = w_2 = \frac{f_{stall}}{b\sigma_{max}} \quad (7)$$

where;

$$\sigma_{max} = n\sigma_{yield} \quad (8)$$

Therefore;

$$l_1 = l_2 = r_{aspect} w_1 \quad (9)$$

3.4.2 SOLUTION OF a_1 AND a_2.

The geometry a_1, l_1, l_2, w_1, w_2, b and Young's Modulus E, will determine the structural loading on the input drive (usually an actuator) in mode 1. This loading must be minimised to optimise efficiency. The only remaining variable to achieve this is **a1**. In addition, the selection of this variable must be tempered with the criterion of stress loading.

The relation between input displacement, input force, input compliance and maximum stress is complex and is determined by the solution of simultaneous equations of the third order.

Figure 2 shows the idealised deformation of the flexors at full piezo extension, but with no output load (mode 1). Although this is simplified, we can calculate the maximum stresses due to bending, which will naturally occur at the periphery of flexors, assuming that the stiffness of the main beam remains far greater than that of the flexors.

We know that the extension of the drive component is given by;

$$x_{i_{true}} = x_{i_{free}} \cdot \frac{s_{rot}}{(s_{rot} + s_d)} \quad (10)$$

where s_{rot} is the input compliance of the structure, and therefore;

$$\theta = \frac{x_{i_{true}}}{a_1} \quad (11)$$

If the beam is sufficiently stiff over the length a_1, we can assume that the top end of the flexors undergo the same angular deflection. Thus;

$$\theta = \frac{x_{i_{true}}}{a_1} = \frac{-Pl_1^2}{2EI_1} + \frac{Ml_1}{EI_1} = \frac{Pl_2^2}{2EI_2} + \frac{(Fa_1 - M)l_2}{EI_2} \quad (12)$$

By a similar argument, the lateral deflection of both flexors must be approximately equal. And so;

$$\frac{Pl_1^3}{3EI_1} - \frac{Ml_1^2}{2EI_1} = \frac{-Pl_2^3}{3EI_2} - \frac{(Fa_1 - M)l_2^2}{2EI_2} \quad (13)$$

These equations can be solved to find F, M and P and therefore the input compliance s_{rot} from;

$$s_{rot} = \frac{x_{i_{true}}}{F} \quad (14)$$

The maximum stresses within the structure can be estimated by considering the individual contributions from direct force **F**, lateral force **P** and bending moment **M**. The peak stress zones are indicated in Figure 2.

Finding a value for a_1 is achieved by testing solutions for increasing values, and terminating when a value is found which results in a good matching efficiency. This matching value is given by;

$$k_{rot} \leq \frac{s_{rot}}{s_d} \quad (15)$$

This solution is then tested for stress safety by re-calculating the newly found input compliance. Re-running the torsion algorithm checks for the maximum stress obtained in mode 1, and increases a_1 until the stress value is acceptable.

The final value of a_1 and the input compliance is used to re-calculate the real input movement, as before, and thence the value of a_2, necessary to generate the required output of the structure. The gain is given by;

$$g = \frac{x_o}{x_{i_{true}}} \quad (16)$$

and therefore;

$$a_2 = a_1(g-1) \quad (17)$$

3.4.3 SOLUTION OF d, e_1, e_2.

To arrive at an appropriate geometry for this section of the structure, a value for the main beam input compliance must be chosen. There is little point in choosing an excessively low value. For this reason, an input compliance for the beam structure is sought, simply supported at both ends, and measured from the top of hinge 2. The value of this is set to;

$$s_{i_{beam}} = k_{beam}(s_d + s_1 + s_2) \quad (18)$$

and is found iteratively by locking $e_1 = .99 d$ and $e_2 = e_1$.

Values of **d** beyond this imply geometries which are unnecessarily bulky. It is effectively the sum of the compliances of the two half-beams of the structure, reflected through appropriate pivotal centres. Each component therefore has a *weighting factor* associated with it, derived from geometrical values. The individual components considered are;

i) Compliances due to linear extension/compression of the hinges (flexors) s_{f1} and s_{f2}.

ii) Compliance due to the bending of the two beam structures to the left and right of the axis of the piezo; s_{b1} and s_{b2}.

(Note that stiffness due to the bending of the hinges is ignored.)

If hinge 2 is in compression and the drive experiences the same force, by assuming rotation about h_1 we can say;

$$S_{of2} = s_{f2}\left(\frac{a_2+a_1}{a_1}\right)^2 \qquad (19)$$

where s_{of2} is the output compliance solely due to hinge 2 compressing. Similarly, by pivoting about hinge 2, we get the effective output compliance due to hinge 1 extension as;

$$S_{of1} = s_{f1}\left(\frac{a_2}{a_1}\right)^2 \qquad (20)$$

The compliances of each hinge s_{f1} and s_{f2} are simply given by;

$$s_{f1} = \frac{l_1}{Ebw_1} \qquad s_{f2} = \frac{l_2}{Ebw_2} \qquad (21)$$

The output compliance of the structure can now be calculated for the case when e_1 and e_2 are nearly d. This compliance is;

$$S_{o_{best}} = (s_1 + s_{b1})\left(\frac{a_2}{a_1}\right)^2 + (s_2 + s_d)\left(\frac{a_1+a_2}{a_1}\right)^2 + s_{b2} \qquad (22)$$

From this point, the e_1 and e_2 values are iteratively decreased to reduce mass, and the output compliance allowed to rise, again according to a programmable tuning parameter, i.e.;

$$\frac{S_o}{S_{o_{best}}} = 1 + k_{beam} \qquad (23)$$

The overall force-displacement efficiency is defined as;

$$\eta_{fd} = \frac{x_{o_{best}} f_{o_{stall}}}{x_{i_{free}} f_{i_{stall}}} \qquad (24)$$

4 A WORKED EXAMPLE.

As an example the program was used to find a solution to the problem of generating a high force output with 100μm displacement, from a Tokin Corp. NLA 2x3x18 Actuator employing a single stage amplifier. The ensuing design was manufactured using an Electric Discharge Machining Process, and assessed for output displacement and output stiffness.

It must be stated that the program does not produce a unique solution for a particular problem since there are many degrees of freedom to solve. Additionally, many design parameters exist which the program calculates but does not solve for, such as device mass.

A piece of 2.5mm thick SAE 4340 steel was chosen as the billet from which the device could be manufactured.

A typical family of solutions is shown in Figure 5. The program control parameter of device efficiency was used to generate this group, and as can be seen, devices approaching 80% efficient can be designed, but with the penalty of increasing device mass. Generally, the resonant frequency and hence speed of response of such a family of devices, can be related to the function;

$$f(m) \propto \frac{1}{\sqrt{m}} \qquad (25)$$

and this relationship is shown in Figure 6.

A compromise between response speed and efficiency would normally be resolved by particular factors in a real design problem. For the purpose of the design study, the 70% efficient device was chosen. A screen-dump of the program, running the example is shown in Figure 7. The geometrical data generated by the beam designer program is shown in Table I.

These values were used to construct a finite element mesh using PIGS 4.2, via the DXF output option of the beam designer. The results obtained from the F.E. analysis are shown in Table II, along with the values predicted by the beam designer program, and measurements taken from the device. Stress measurements on the device itself were not possible.

Most parameters from either the program or the real device fall within a 12% band of those generated by F.E. analysis. The stress values obtained by the beam designer program were all conservative by this standard, i.e. the actual stresses predicted by the F.E. technique were slightly lower than those predicted by the program. The discrepancies produced are believed by the authors to be mostly due to inadequacies in the modelling which fail to take account of *local* distortions and stresses generated in the beams by the hinges, for example, within the hinge fillets. The `beam designer' modelling also fails to account for distortions of the host structure at the `static' end of the actuator. Most seriously, the F.E.A. prediction of output displacement is poor by comparison with the `beam designer'.

5 CONCLUSIONS.

A methodology has been offered for the design of simple linear displacement amplifiers for use with piezoelectric multilayer actuators. Finite Element Analysis of a structure proposed by the designer program indicated an acceptable degree of correlation between the F.E.A. results, the performance and stressing data generated by the program, and real data obtained by experimental measurement of the real device.

Theoretical results show that highly efficient designs (80%+) are possible, but result in bulky structures. The relationship between device efficiency and device mass is one of diminishing returns, favouring compromises in the region of 50% to 70% force displacement efficiency. This results in device geometry of comparable size to the prime mover.

6 FIGURES.

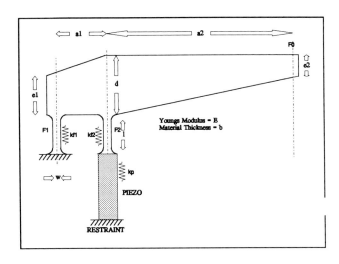

Fig 1 One type of displacement amplifier using flexural hinges

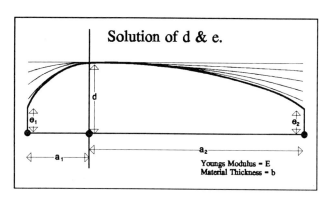

Fig 4 Iterative solution of d, e_1, and e_2 values

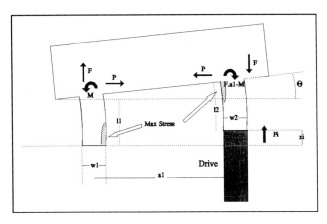

Fig 2 Solution of hinge separation

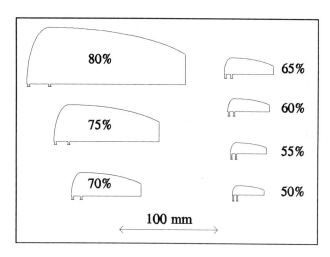

Fig 5 Family of solutions for the 100μm amplifier. Percentages shown refer to structural efficiency

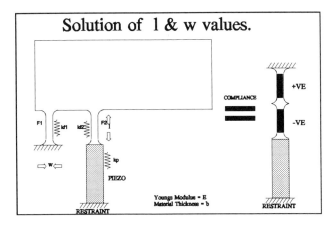

Fig 3 Solution of the hinge dimensions

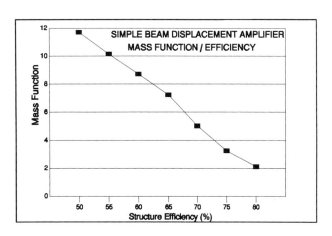

Fig 6 Mass function against efficiency for the 100μm amplifier

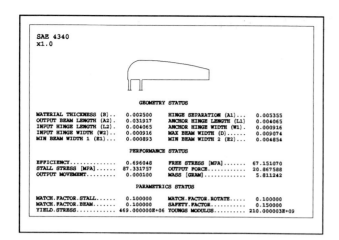

Fig 7 Screen-dump of the beam designer program

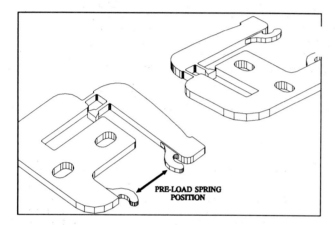

Fig 8 Two views of the displacement amplifier

DIMENSION	VALUE(mm)
a_1	5.36
a_2	31.90
l_1	4.07
l_2	4.07
w_1	0.92
w_2	0.92
d	9.07
e_1	0.89
e_2	4.85
b	2.50

Table 1 Geometrical data produced by the Beam Designer Program

PARAM'	DESIGN	F.E.A.	REAL
Displacement	100 μm	92 μm	101 μm
Stall Force	20.8 N	18.6 N	18.2 N
Stall Stress	87 MPa	72 MPa	-----
Free Stress	67 MPa	59 MPa	-----
Efficiency	69 %	57 %	61%

Table 2 Comparision of Modelling Performance with F.E.A. and a Manufactured Device

Accelerometers for suspension control

P KELLETT, BSc, CEng, FIEE
Lamerholm Fleming Ltd, UK

Synopsis
Different types of accelerometers, specially developed for the automotive industry, are described, notably piezo-electric, strain gauge and force balance devices. All these devices have been proved in the field. Two systems of "adaptive suspension" illustrate how these accelerometers have made possible the production of economic systems. The high performance required from the accelerometers has only been achieved by the close integration with the processor chip. The technology has made possible the introduction of adaptive suspension, previously only available in "executive class" vehicles, to mass produced low cost cars.

Introduction

In order to control the suspension of a vehicle it is necessary to measure and possibly predict the movement of the vehicle relative to the road. Accelerometers have been found most useful in this context. Such sensors have been used for laboratory measurement for many years but have been unsuitable for use in automotive applications because of their high cost.

In suspension control there are two types of accelerometer requirement:-

(a) Frequency response extending down to less than 1Hz.

(b) Frequency response extending down to "DC" or steady state acceleration.

The first requirement can be met by a piezo electric type of accelerometer and the second either by a strain gauge bridge form of construction or by a "Force Balance" or "servo" accelerometer.

Piezo Accelerometer.

In a typical piezo accelerometer the performance at low frequencies is compromised by the presence of the associated pyro electric effect. In most previous applications of the piezo effect to transducers the thermal time constants of the structure have been so much greater than the electrical time constants associated with the frequencies being processed, that this pyro electric effect has been negligible. In dealing with these very low frequencies, associated with suspension systems, this is no longer the case and conventional piezo electric sensor construction has been found unsatisfactory.

In Fig.1. a form of construction is shown in which two discs of Piezo material bonded on either side of a thin, thermally conducting plate, are arranged and connected in such a way that the piezo electrical signals from the two elements due to axial acceleration are additive while the pyro electric signals cancel each other.

This technology has now been proved in production vehicles where such accelerometers have been used to measure the vertical movement of the vehicle body in order that some control may be applied. In the first application of this technique, the accelerometers were integrated into the controller and assembled on the PCB.

The later versions, which have

now been developed, are "stand alone" devices which incorporate appropriate customised band pass filtering to meet the differing requirements of each application. Two typical response curves are shown in fig.2 which illustrate how the response may be tailored to meet differing requirements.

In the automotive industry cost is of great importance. In this instance a target price of 10$ is achievable subject to a high volume requirement.

Strain Gauge Accelerometers.

Strain gauge accelerometers usually consist of a mass attached to a cantilever or similar arranged in such a way that when the structure is subjected to acceleration a force is produced acting on the mass which in turn results in a strain being induced in the cantilever. Strain gauges, arranged in the well known bridge configuration are used to measure this strain. Such devices suffer from problems associated with Zero offset due to the unbalance of the bridge under zero input conditions. This situation is exacerbated by temperature effects which result in the offset varying with temperature. The existence of the offset reduces the ability of the system to measure small signals due to low accelerations. This may have the effect, for example, of increasing the time delay for the system to respond to the commencement of a vehicle manoeuvre.

On a laboratory or instrumentation scale it is possible to provide trimmer controls which can be adjusted to minimise this effect. For automotive applications this approach is too expensive to consider. A very simple alternative approach has been used with satisfactory results.

It was possible, in this application, to mount the accelerometers in the same box as the controller. A stable temperature sensing device is also mounted in close proximity to the accelerometers. During testing and "burn in" it is possible to measure the offset of each sensor over the temperature range, and to store the calibration of each sensor in the non volatile memory of the processor in the controller.

Thus in use the microprocessor is able to compare the instantaneous value of the output of each sensor with its stored associated calibration at the temperature prevailing at the time of the measurement. This has been found entirely satisfactory in practice and in association with the previously described piezo sensor has been used for over three years in an adaptive suspension system in a luxury car in full production.

Servo Accelerometer

Although the method of compensating for the zero offset drift of strain gauge accelerometers, as described above, is practical and proven, a sensor with no inherent zero offset would be better. Such a sensor can be made and is generally known as a servo, or force balance accelerometer.

Fig. 3. shows in schematic form, such a sensor. In this diagram a mass 10 is supported on the free end of a beam or cantilever 1 pivoted at point 2. The position of the cantilever is detected by the electo-optical arrangement of 8 and 9. The cantilever will tend to move in the direction A/B under the influence of an acceleration on that axis. A restoring force is applied through the coil and magnet arrangement of 5 and 6. The acceleration is proportional to the current in coil 6 required to restore the cantilever to its zero position.

Fig.4. shows the system circuit schematic associated with the sensor. In a typical application the microprocessor 12 is the control processor for the suspension and it has sufficient capacity to manage a number of associated sensors. In this particular embodiment of the basic principle, it has been found possible to achieve +/- 11 bits of effective resolution at a cost which is compatible with automotive sensor targets.

Typically, a full scale sensitivity of the order of 1g is required for these suspension control accelerometers. The servo accelerometer described has an adequate performance as can be seen in the performance curves of Fig.5. and Fig. 6. The zero offset and its drift with temperature, over the full temperature range of -40C to +85C, are better than many expensive instrumentation type sensors. The total drift span of only 2% of full scale compares very favourably with the 5% to 10% typically specified, over a more restricted temperature range, for piezoresistive accelerometers.

The extremely low variation in sesitivity over the wide operating temperature range, shown in fig.6. is also better, by a factor of 10, than that obtainable with many instrumentation type sensors. It is important to remember that these variations of only one or two percent over a very wide temperature range are also linked to a high sensitivity of the order of 1g full scale.

In high volumes the 10$ target is again achievable.

In practice, because of the close interaction between the sensor and the processor it is possible to effect the necessary filtering digitally and indeed to make the filters programmable. The sensor, due to its own inherent processing power, becomes capable of communicating with other processors and of adapting its own characteristics should it be required to do so. In the suspension controller to be described later, a number of sensors share one processor which still has enough spare capacity to perform the suspension control function.

Suspension Control Systems using the accelerometers described previously.

In order to isolate the passengers of a wheeled vehicle from vibration caused by the passage of the vehicle over irregularities in terrain over which it is travelling it is necessary to introduce some mechanical compliance between the wheels and the vehicle body. If a spring based suspension is employed then some form of damping is required to restrict the movement of the vehicle relative to the wheels. Almost all modern vehicles use a damped spring suspension system to support the body and to isolate it from wheel vibrations. Although a wide variety of spring and damping techniques are in use the essential principle is the same.

In order to make a satisfactory spring and damper system it is necessary to find acceptable compromise solutions to several conflicting demands. In order to minimise roll, squat and dive very stiff springs are required, soft springs are needed to provide good ride comfort on smooth road surfaces.

A small number of mainly prototype vehicles have been built where the spring and damper arrangement has been replaced by hydraulic rams connecting the vehicle body to the wheels. The required isolation is achieved by adjusting the drive to the hydraulic rams. The effective spring rate and damping of the suspension system may be entirely determined by the control system driving the rams. Whilst such a system is capable of providing significant performance advantages, current high costs preclude its wide spread adoption. These, fully active suspension systems, show great promise once the cost problems of the suspension components can be solved.

It has been determined both theoretically and experimentally, that an improvement in suspension performance may be obtained by changing the damping characteristics in response to driver and vehicle behaviour. There are now several vehicles in production where the state of two or three step dampers is selected by an electronic control system with or without manual override. These systems are usually referred to as adaptive suspension systems. Depending on the type of vehicle and the choice and design of the system components, including the control algorithms, these systems have improved the overall suspension performance by differing amounts. In the best cases very significant improvements have been achieved.

The common feature of most production installations has been the very considerable increase in cost compared to a traditional system. In some cases the benefit obtained has been worth the increase in cost but in many, the benefit has been considered to represent poor value for money, either by end users when it is offered as an option or by marketing departments at the product planning stage.

Several projects have suffered from the widespread use of top range models as the launching pads for new concepts such as adaptive suspension. Most large vehicles offer an acceptable compromise suspension performance, achieved by traditional means. The simple addition of an adaptive system provides little benefit unless it is sophisticated and complex and therefore expensive. The application of adaptive suspension to lower cost vehicles has frequently been assessed on the basis of work done on "flagship" projects and has foundered at an early stage due to the excessive cost. Nevertheless some of these "flag ship" projects have been very satisfactory and have provided benefits which cannot be achieved by conventional means. The systems used by Rolls Royce and Bentley, based on the accelerometers described above are instanced as examples of such successful systems.

The RD195 Adaptive Suspension Controller has been designed to counter these problems and to pave the way for controlled suspension systems in small lower cost vehicles.

The Adaptive Suspension Controller.

The major objective of this design exercise was to engineer a practical system using wherever possible proven technology, to produce a low cost system applicable to the smaller cheaper vehicle.

To this end the system was based on the use of two step solenoid controlled dampers. In order to minimise the cost of both the controller and the installation in the vehicle three integral accelerometers are mounted in the controller. A signal from the existing vehicle speed instrument is also used. Two further inputs are provided to access driver intent - a mode select switch and a connection to the brake pedal switch.

There are two outputs from the controller - a power output to switch the damper solenoids and a fault lamp to warn the driver of any suspension system faults detected by the controller.

The various inputs are measured at regular intervals and the data acquired is processed by a high speed processor according to a set of rules. The dampers are set to the hard or soft setting according to the results of these calculations.

The high accuracy and low threshold of the servo accelerometers has made possible a system with out a steering sensor. This represents the elimination of a relatively expensive sensor which also has a high installation and maintenance cost.

Control Strategy

The basic control strategy sets the hard setting as the default state. Each acceleration input is compared with a threshold value and the hard state is requested as long as this state of affairs persists. When the accelerations are such that the signals fall below the threshold the soft setting is requested.

The thresholds are mapped against speed and some 30 constants are provided for each of the accelerometer inputs.

The flexibility of the system is such that it is possible to ensure that , for example, at very low speeds hard is always selected. A rather more sophisticated technique is used to improve handling under fast lane changing conditions. This simply stated, allows for a more rapid response to lateral accelerations, characteristic of lane changing.

Other strategies, applicable to the particular type of vehicle can easily be accommodated by the "constant tuning" process. In order to facilitate this "tuning" process a comprehensive PC based control software suite has been developed which allows real time control of the operating RD195 controller on a working vehicle, data acquisition from a running RD195 controller, constant map editing and simulation of the RD195 using either acquired or synthesised data. Connection is made by connecting the RD195 to the serial port of the PC. and by accessing the user friendly software package.

Conclusion

This paper has described a number of sensors developed to meet the particular requirements of the automotive industry and has also attempted to describe how these sensors can be and have been used to construct practical low cost suspension control systems.

Patents Ref:

European 0374159 Accelerometers and Associated Control Circuits.

PCT/ WO 8903019 Electronic Controller Unit for correction of sensing errors.

PCT/ WO/91/09315 Signal Processing Circuits.

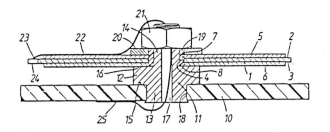

Fig 1

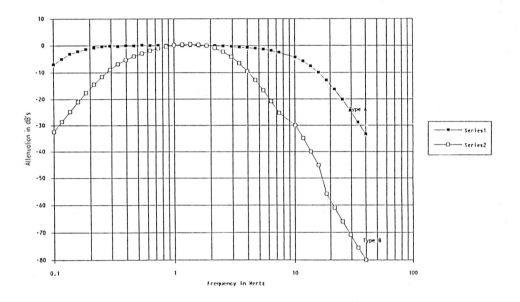

Fig 2 Frequency response of piezo accelerometers

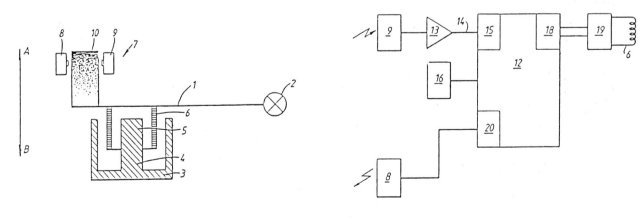

Fig 3

Fig 4

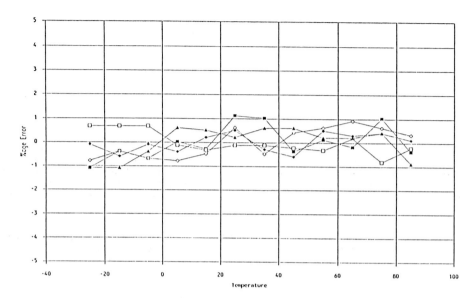

Fig 5 Zero offset drift of servo accelerometer

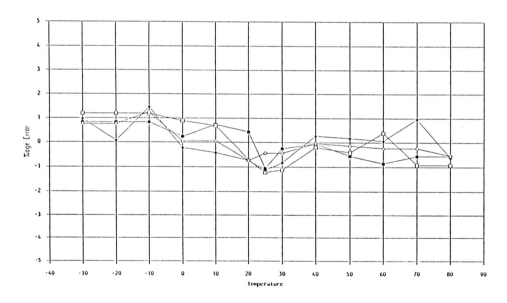

Fig 6 Servo accelerometer full scale sensitivity variation over temperature

A rapid response thermal tactile sensor

G J MONKMAN, BA, BSc, MSc, PhD, MIEEE
University of Hull, UK

SYNOPSIS: Tactile sensing is one of the major emerging technologies now starting to be seriously considered by the robotics community, for industrial use. Whereas previously, tactile sensors have existed mainly in the form of research prototypes, industrially viable devices are now in use. Physical pressure sensor arrays which no longer suffer from the drawbacks of poor repeatability, durability, excessive hysteresis etc, are now commercially available. However, mechanical strain is only one aspect of the very rich environment of measurands available to potential contact sensory devices.

Thermal tactile sensing not only provides the usual geometrical data of an objects physical topology but also has the additional advantage of yielding information on the objects physical construction. Unfortunately, thermal techniques have hitherto suffered from the problem of slow temporal response and recovery compared to other tactile sensing systems.

This paper outlines a new method of thermal tactile sensing which combines mechanical, electrical and optical technology in order to achieve response times between one and two orders of magnitude faster than conventional thermal contact measurement techniques.

1. INTRODUCTION.

According to Harmon [1] the requirement for a tactile sensor to emulate the human finger is a dynamic pressure range of three orders of magnitude with a resolution of 20 by 20 tactels per finger. A spatial resolution of better than 2mm and a time response within 10 mS with low hysteresis are also desirable. Such a time response may be a little optimistic, even in the case of a physical displacement transducer this represents a displacement velocity of 0.1 m/s for a 1 mm depression. Though perhaps conservative for most robot transitions, this type of velocity is not very representitive of fine grasping movements. Its main significance lies in active sensing where the sensor is made, usually by movement over a surface, to create the tactile sense signal.

As one may observe when touching objects with the finger tips, temperature is also an integral part of the human overall tactile sensing strategy. Thermal effects also help determine the 'feel' or apparent texture of a surface.

2. THE HUMAN TACTILE SENSOR.

We cannot consider the finger tips as tactile sensors in isolation without due regard to the human entity as a whole. As pointed out by Lederman [2], in everyday life our fingers explore their environment, actively pushing against objects to determine their form. Unlike many inanimate compliant membranes, the very nature of flesh allows it to return rapidly to its original profile after depression and release. Constant blood flow, and other physical movements, help to augment this positive recovery effect. Add to this vision and other sense parameters, together with an intelligent closed control loop between the hands and all the senses, and it is not surprising that less than perfect tactile sensing can be tolerated. In fact, the human fingers contain at least 5 different receptor types, many of which are connected in a one-to-many configuration [3]. This must inevitably result in a high degree of cross-talk.

Surface texture and the relative thermal conductivity of a material accounts for much of the confusion the human senses are subject to. Whenever an unknown object is encountered, the fingers are moved over the surface to give a feel for its surface profile, material stiffness, temperature etc. Unfortunately the temperature sensing capability of the human machine can be very misleading. A simple test of this consists of estimating, by touch, the relative temperature of a selection of objects before measuring their actual temperature with an accurate thermometer or thermocouple probe. This kind of experiment will be familiar to most children who have dipped a finger into bowls of water of varying temperature in an attempt to estimate their relative temperatures.

What our fingers detect thermally is not the absolute temperature of a material alone, but also its thermal conductivity and diffusivity. This is because the human finger is a source of heat, which at approximately 34°C is slightly above the ambient temperature of most objects we encounter. What the nerves in the fingers detect is the outflow of heat from this source. This suggests that the human temperature sense may be emulated by thermal conduction sensing.

3. THERMAL SENSING.

In the same way that piezoelectric materials exhibit an electrical output when subjected to a mechanical stress or pressure, so do pyroelectric materials when heated or cooled. In fact the polyvinylidene flouride used by Dario [4] for piezoelectric tactile sensors also exhibits a pyroelectric effect [5].

Most thermal devices for tactile use are active. They consist of both a heat source and sensor rather than attempting to measure absolute temperature by means of the sensor alone. Russell [6] demonstrates a 10 by 10 array of tactels using integral heating elements and thermistors. An extension to this is the 4 by 4 thermal array superimposed on an 8 by 8 capacitive array of force sensors shown by Siegel and co-workers [7].

In both the above cases some degree of material recognition is possible due to the different thermal conductivities of dissimilar materials. Unfortunately, in both cases response was found to be very slow for the thermal sensors. A 90% rise and decay time of roughly 18 seconds was experienced by Siegel, whereas Russell appears to have achieved a 4 second rise time but again a decay of around 20 seconds. A device which yields superior performance in this respect is now described and compared with a commercially available force sensing device.

3.1 Pyrometers for Thermal Tactile Sensing.

Very few semiconductor materials perform well in the far infrared. Mercury/Cadmium Telluride (Hg/Cd Te) and Lead/Tin Telluride (Pb/Sn Te) are active between 8 and 14 μm but require an operating temperature of $77\,°K$ and are therfore usually restricted to sophisticated military applications [8].

More practical are ferro-electric crystals, some of which are capable of very fast responses, so much so that they can be used to detect infrared signals modulated at several MHz. A typical arrangement is shown in figure 1 [9].

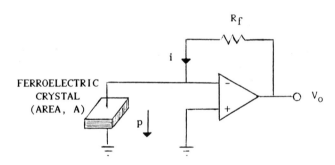

Fig 1   Basic infrared detector circuit

Analysis of this circuit gives:

$$V_0 = R_f\, i = R_f\, A\, p\, \frac{dT}{dt} \qquad (1)$$

where p is the pyroelectric coefficient (C m^{-2} K^{-1}), dependant on the dielectric properties and specific heat of the crystal element [10]. The factor R_v is the responsivity of the crystal material and depends on both the thermal response to incident radiation and the pyroelectric response to the temperature change $\triangle T$ due to incident radiation of power W(t).

$$c\,\frac{dT}{dt} + G\,\triangle T = e\, W(t) \qquad (2)$$

where c is the heat capacity of the crystal (J K^{-1}), G is thermal conductance (J s^{-1} K^{-1}) and e is the fraction of effective incident power thermalising the crystal.

The voltage responsivity R_v in volts/watt of incident radiation is given in {3}, where R is the electrical input resistance.

$$\text{Voltage responsivity}\quad R_v = \frac{A\, e\, R\, p}{c} \qquad (3)$$

Burfoot & Taylor [10] (among others) provide tables of values of R_v and other parameters for the selection of suitable materials. In most cases thermal relaxation time restricts the frequency response to the range 1 to 100 Hz [9]. Bandwidth of the detector can be increased by decreasing the input resistance R, but as can be seen from {3} this will result in a corresponding loss of responsivity.

Re-arranging {3} gives:

$$p = \frac{R_v\, c}{A\, e\, R} \qquad (4)$$

Combining {1} and {4} produces an expression {5} for the output voltage.

$$V_0 = \frac{R_f}{R}\,\frac{R_v\, c}{e}\,\frac{dT}{dt} \qquad (5)$$

From {5} it can be seen that improving the responsivity will increase the output voltage V_0 but, as mentioned previously, at the cost of a reduction in operating bandwidth. A more practical means of raising the output voltage is by increasing the gain factor R_f/R within the usual gain bandwidth product and noise constraints of the chosen operational amplifier.

What is important about {1} is the influence of $\partial T/\partial t$ on the output voltage V_0. This shows that only a change in temperature will produce a resultant output V_0. The limiting factor with regard to response time is due to the crystal resistance and capacitance together with that of the housing (which is usually dominant). As most commercially produced devices are only available in hermetically sealed enclosures (often because of the deliquescent nature of many ferroelectric crystal materials), there is little the end user can do to improve the behaviour. However, during fabrication of large arrays some steps may be taken to optimise on the desired parameters, in this case response time.

The upper limit on operating temperature for most ferroelectric materials is in the region of $60\,°C$. This may pose a problem in a few robotic applications where relatively high temperatures are involved, though this is likely to be a minority of cases.

Such devices are commonly available in the form of pyrometers of the type used in infrared security detector systems. The sensor elements have a typical radiation detection frequency range of 5 to 15 μm [11]. Being electrically polarized, usually 2 complimentarily connected elements are included in each pyrometer package. This accounts for the opposite polarity in the contact and removal response curves of figure 3. Unfortunately, these are essentially passive devices which provide no heat source of their own. Consequently an additional heat source, for example a simple resistive heating element, must be provided. Figure 2 shows a typical construction of such a thermal tactile sensing device. Though the actual elements are very small (approximately 2 mm square) they are normally housed in a TO39, or in this case, a 6 pin DIL package about 7mm square.

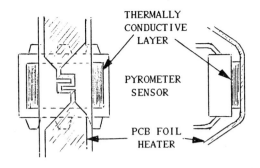

Fig 2 Pyrometer Tactel

In the above experimental device, the heating element consists of a simple electrically resistive track etched on to a layer of PCB foil, driven by a current of approximately 2 amps. The thermally conductive layer comprises of an opaque aluminium film to elliminate effects of electromagnetic infrared radiation whilst at the same time allowing heat to be transferred though the film and radiated on to the sensing elements. It is necessary for this film to be as thin as possible to prevent it from acting as a heat sink which would reduce the response time.

Figure 3 shows a typical response curve for contact and release from a thermally conducting subject, as would be observed from the output of an amplifier such as that in figure 1 or 4. As can clearly be seen from these plots, both the rise and decay times are relatively fast. This is particularly apparent in the 63% rise time of approximately 260 mS (600 mS for 90% rise time) for the response caused by initial contact. The recovery to steady state is somewhat slower at 800 mS (1.8 seconds for 90% decay time). However, with the sensor elements being basically capacitive in nature, these sensors only detect temperature change and cannot measure absolute temperature levels. This is of course very similar to the human tactile model. The second pulse, of opposite polarity, represents the effects of subject removal and has a similar time response to the contact curve. When used with thermally insulating subjects, the pulses are in the reverse direction, due to the increase in energy as a result of reflected heat.

A suitable amplifier circuit, designed to obtain a good frequency response from the pyrometer sensor, is shown in figure 4. The design criteria is not strict and may be adapted to suite the intended use.

One of the main advantages of this kind of device is its inherent ability to be fabricated into arrays in VLSI form. Under these circumstances, the foil heaters may be replaced by small semiconductor devices (zener diodes, peltier devices etc.) fabricated into the same VLSI array.

Investigations into the pyroelectric properties of Langmuir-Blodgett films in the UK shows some promise for use in thermal sensing [12]. Though they lack the sensitivity of ferroelectric crystals (a pyroelectric coefficient of about 2 compared to up to 76 for Lead Lanthanum Zirconate Titanate [9]), their inherent ability to be formed and cut into arrays provides some distinct advantages. Due to the measurement of heat flux changes, as opposed to temperature difference, ferroelectric crystal pyrometers are roughly three orders of magnitude more sensitive than thermistors [13]. Hence, when used in tactile sensors, an improvement in response time at the expense of reduced sensitivity may be justifiable.

Ferroelectric devices can also be electrically modulated to produce a time varying thermal emission [10]. It is conceiveable that such an alternating heat source in conjunction with a similar detector device could enable continuous measurements of thermal conduction to be made. Given a fast enough response time of the thermal emitter, it is possible to pulse the emitter and observe changes in the reflected thermal energy. This would naturally be greater when the device is in contact with a thermally insulating, rather than conducting subject.

Though the simple heating elements used in the prototype of figure 2 are likely to be too slow, ferroelectric emitters, or heating elements of the kind used in dye diffusion thermal colour printing heads are much faster. Such heating elements are deposited as linear arrays using thin film processes and are capable of rise times of around 10 mS [14]. For purposes of signal to noise ratio improvement Nickel-Chromium heaters have actually been integrated into pyroelectric infra-red detectors [15].

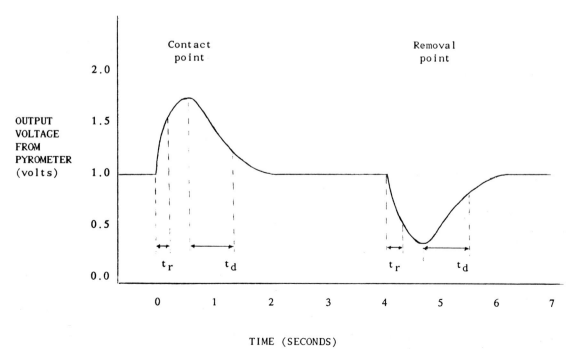

Fig 3 Pyrometer response curves

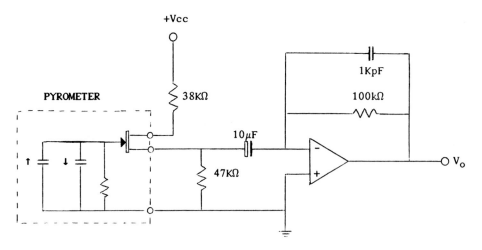

Fig 4 Pyrometer amplifier circuit

3.2 Comparison with Force Sensing Tactel.

Many force sensing techniques have been experimented with in the past including the reflection of light from a deformable membrane [16], megneto-resistive [17] and capacitive [18] devices, to mention but a few. In almost all cases the limiting factors in terms of repeatability, hysteresis and response time are determined by the mechanical properties of moving parts, in most cases flexible membranes.

Piezoresistance & electrical conductivity is the technique by which a compliant skin, usually made from some form of conducting rubber or carbon impregnated foam, changes its electrical resistance when subjected to mechanical pressure leading to elastic deformation. The relative simplicity of this method has made it one of the most researched areas of tactile sensing technology. A 16 by 16 element tactile sensor capable of 256 pressure levels per element has been demonstrated [19]. This measures the change in conductivity of a rubber skin. A response time of 75 mS is claimed, and some degree of hysteresis is inevitable. Sensor drift is also a problem.

One of the earliest commercially available tactile sensor arrays, manufactured by GEC, used a compliant piezoresistive mat sandwitched between two layers of parallel conductive strips. These conductors may be scanned electrically to give individual tactel pressure measurements at the points where two strips overlap. In this case the mat was formed from a fine yarn impregnated with carbon [20]. Similar tactile sensors using carbon fibre as a flexible conductor have been proposed [21], [22], [23]. This method goes some way to solving the problems of metal fatigue experienced when using copper conductors on flexible PCB substrates. In order to mitigate the effects of poor contact resistance repeatability Helsel and colleagues [24] employ an electrically conducting fluid medium between the electrodes, enclosed by a latex membrane. The natural properties of a fluid also enable rapid and repeatable recovery of the structure.

One device which does not appear to suffer the problems of severe hysteresis and poor longevity, traditionally associated with mechanical tactile sensors, is the FSR™ (force sensing resistor) made by *Interlink* [25]. These devices, made from conductive polymer, exhibit less than 2 mS response times and are almost totally imune to temperature variations. Their use as robotic tactile sensors, either individually or in arrays, has been dealt with extensively by Speeter [26]. Therefore this work will concentrate on their use in conjunction with other forms of tactile sensor rather than on the characteristics of the FSR™.

When solid objects are encountered, the displacement of the FSR™ tactel is extremely small, hence the electrical resistance change experienced is proportional to the applied force. Therefore the measurand may be considered to be stress rather than strain. On the other hand, when the tactel comes into contact with soft or compliant surfaces, considerable displacement is possible from the initial contact point revealed by, say thermal and/or resistive sensors. This displacement, measured by the robots position encoders, represents the strain undergone by the object. In conjunction with the force measurement from the FSR™ it can be used to calculate the elastic modulus of the object material. In many cases compliance must be part of the tactile sensor construction itself [27], particularly if the tactile sensor housing constitutes part of the end-effectors gripping surface.

$$\text{Elastic modulus, } E = \frac{\text{stress}}{\text{strain}} = \frac{\sigma}{\epsilon} \qquad (6)$$

$$\text{where stress,} \quad \sigma = \text{force}/\text{area} \qquad (7)$$

and, the elastic displacement from an initial state ℓ_0 to a final state ℓ gives the true strain ϵ:

$$\epsilon = \int_{\ell_0}^{\ell} \frac{d\ell}{\ell} = \ln\left(\frac{\ell}{\ell_0}\right) \qquad (8)$$

The actual area of contact may be assumed constant for a small tactile element so that the final measurand may either be stress, strain or elastic modulus depending of whether the tactel is in contact with a rigid or non-rigid surface.

Clearly, it is not an easy task to compare two sensors which are normally employed for the measurement of two totally different, and somewhat mutually independant, parameters. However, their characteristics complement rather than conflict when both devices are combined and used for the identification of subject material properties.

Like the anthropormorphic model, combinations of several tactile sensors has been investigated. Siegel combined both thermal and mechanical sensors [7]. Unfortunately, as mentioned previously, the thermal sensor had a much poorer temporal response than the mechanical tactile device. Figure 5 shows a combined thermal and mechanical device with the addition of an electrical resistance sensor. This allows a subjects electrical conductivity to be ascertained (something the human tactile sense cannot do). The thermal sensor being a pyrometer and heater, of the type depicted in figure 2, the response is comparatively fast.

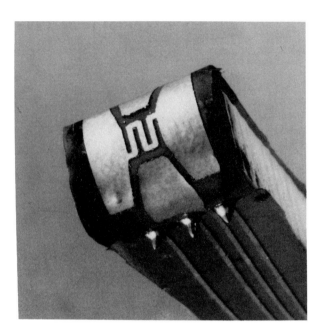

Fig 5 Combined thermal, electrical and mechanical tactile sensor

The electrical resistance measurement is provided by means of additional sensor tracks etched onto the same PCB foil as the heater element. This allows the sensor to distinguish between good electrical and thermal conduction properties. There are very good reasons for this. Carbon fibre is a reasonable electrical conductor, though a relatively poor thermal conductor. Conversely, mica and similar materials, used as electrical barriers between high power dissipation electronic components and heat sinks, are extremely good electrical insulators but also good thermal conductors. The addition of force sensing is incorporated at a convenient position beneath the thermal sensor. As previously discussed, pressure sensitive devices are commercially available, some of which are fabricated as flexible films such as force sensing resistors [25].

4. CONCLUSIONS.

This paper has considered the human tactile model together with its failings when used to measure any single parameter independant of others which may also be present. The necessity to integrate simultaneously, yet independantly operating, sensor elements has been stressed.

A new thermal tactile method has been introduced which has led to an improvement in response time in excess of an order of magnitude over previous thermal tactile sensing means. Combined with other semiconductor sensors, and in the light of current VLSI fabrication technology, the manufacture of tactile arrays using these techniques should prove considerably easier than with many of the purely mechanical tactile methods used in the past. Consideration has also been given to new pyroelectric materials, ideally suited to use in thermal tactile sensors, which offer the possibility of even faster response times.

This new thermal technique has been combined with two other tactile sensing methods to provide a three parameter tactel. Unlike the human model, each measurand enjoys a greater degree of independance with this combination, allowing an estimation to be made of the material properties of any surface which the tactile sensor trio encounters.

REFERENCES.

1. HARMON, L.D. - Automated tactile sensing - *International Journal of Robotics Research* - Vol 1, No. 2, pp 3-32, 1982.

2. LEDERMAN, S. - Human Haptics - *Office of Naval Research Workshop on Dextrous Manipulation and Teleoperation* - University of Oxford, August 1989.

3. ALBUS, J.S. - Brains, behaviour and robotics - *BYTE* - pp39-40, Byte Publications Inc, 1981.

4. DARIO, P., C. Domenici, R. Bardelli, D. De Rossi & P.C. Pinotti - Piezoelectric polymers: New sensor materials for robotic applications - *13th International Symposium on Industrial Robots* - pp 14.34-14.49, Chicago, April 1983.

5. CALDWELL, D.G. - *Novel sensor and gripper design* - BSc Thesis, University of Hull, 1986.

6. RUSSELL, R.A. - Thermal Sensor for Object Shape and Material Constitution - *Robotica* - pp31-4, Vol6, March 1988.

7. SIEGEL, D., I. Garabieta & J.M. Hollerbach - An Integrated Tactile and Thermal Sensor - *Proc. IEEE Intl. Conf. on Robotica and Automation* - 1986.

8. BAR-LEV, A. - *Semiconductors and Electronic Devices* - Prentice Hall, pp 141-142, 1979]

9. LINES, M.E. & A.M. Glass - *Principles & Applications of Ferroelectrics and Related Materials* - Clarendon, pp 141-144, 1977.

10. BURFOOT, J.C. & G.W. Taylor - *Polar Dielectrics & their Applications* - MacMillan, pp 381-384, 1979].

11. RS Components ltd., - Pyroelectric Detector Kit - *RS Data Library* - 10906, November 1990.

12. RICHARDSON. T., G.G. Roberts, S. Holder & D. Lacey - The synthesis and evaluation of novel polysiloxane Langmuir-Blodgett films. - *Thin Solid Films* - Vol 210/211, pp 299-302, 1992.

13. JANATA. J. - *Principles of Chemical Sensors* - Plenum Press, p50, 1989.

14. WEBB, P.W. & R.A. Hann – Measurement of Thermal Transients in a Thermal Print Head used for Dye Diffusion Colour Printing – *IEE Proceedings A; Science, Measurement & Technology* – pp 98–100, vol 138, No. 1, January 1991.

15. ZEMEL. J.N. – *Solid State Chemical Sensors* – Academic Press, 1985.

16. TANIE. K., K. Komoriya, M. Kaneko, S. Tachi & A. Fujikawa – A high resolution tactile sensor – 4th *International Conference on Robot Vision and Sensory Controls* – pp 241–250, IFS, London, October 1984.

17. VRANISH. J.M. – Magnetoresistive skin for robots – 4th *International Conference on Robot Vision and Sensory Controls* – pp 241–250, IFS, London, October 1984.

18. BOIE. R.A. – Cpacitive Impedance Readout Tactile Image Sensor – *IEEE International Conf. on Robotics* – pp370–378, Atlanta, March 1984.

19. VAN BRUSSEL. H. & H. Belien – A high resolution tactile sensor for part recognition – 6th *International Conference on Robot Vision and Sensory Controls* – pp 49–59, IFS, 1986.

20. ROBERTSON. B.E. & A.J. Walkden – Tactile sensor system for robotics – 3rd *International Conference on Robot Vision and Sensory Controls* – SPIE, vol 449, pp572–577, 1983.

21. TROUNOV. A.N. – Application of sensory modules for adaptive robots – 4th *International Conference on Robot Vision and Sensory Controls* – pp 241–250, IFS, London, October 1984.

22. DAVIES. M. – Carbon fibre sensors – 4th *International Conference on Robot Vision and Sensory Controls* – pp 241–250, IFS, London, October 1984.

23. MOORE. T.N., J. Jeswiet, N. Nshama & R.L. Ten Grotenhuis – Development of advanced sensors for industrial robots – 2nd *Engineering Conf. on Computer Aided Production Engineering.* – pp 79–84, 1987.

24. HELSEL. M., J.N. Zemel & V. Dominko – An Impedance Tomographic Tactile Sensor – *Sensors and Actuators* – Elsevier, No. 14, pp 93–98, 1988

25. INTERLINK – *Force and Position Sensing Resistors: An Emerging Technology* – Technical Overview Rev. 2/90, Interlink Electronics, Luxembourg, 1990.

26. SPEETER. T.H. – A Tactile Sensing System for Robotic Manipulation – *The International Journal of Robotics Research* – MIT, Vol9, No.6, pp 25–36, December 1990.

27. RUSSELL. R.A. – *Robot Tactile Sensing* – Prentice–Hall, 1990.

A smart sensor for precision position measurement

C BUTLER, BSc, MSc, PhD, **Q YANG,** DipMtech, PhD
Brunel University, UK

SYNOPSIS A smart position sensor has been developed which is intended for use on coordinate measuring machines (CMMs), machine tools and robots. Existing sensors allow for three orthogonal directions of movement and employ high precision transducers to derive X,Y,Z coordinates. The mechanical complexity results in devices which are extremely effective but are undesirably bulky and expensive to manufacture. The new sensor employs mechatronic principles resulting in a simple and compact mechanical arrangement in conjunction with a novel opto-electronic system working in spherical polar coordinates. The primary sensor consists of an array of optical fibres, a concave mirror mounted on a stylus which contacts the object to be measured. The light intensities detected by the array are modulated by the position and the orientation of the mirror, which changes with the position of the stylus. In the signal processing unit, the detected light intensities are converted to voltage signals, amplified, filtered, and converted to 14 bit digital signals. The incorporated single chip microprocessor finally analyzes the signals and computes the measurement results using a hybrid method based on both empirical formulae and a look up table obtained from the calibration of the sensor. The microprocessor also provides the sensor with the local intelligence of offset control and gain control, trigger threshold adjustment, data communication, together with an interactive operation environment through a LCD and a keypad.

NOTATION

σ standard deviation
SNR   signal-to-noise ratio

1 INTRODUCTION

The issue of precision position sensing and measurement has increasingly become important in the context of manufacturers' continuing pursuit of automation and higher precision manufacture. They are generally required on CMMs [1], machine tools and robots, for example. In practice, contact 3D precision position measurements are preferred in many applications due to their high accuracy and flexibility. There are a few 3D contact position sensors or probes as they are usually referred to in the case of CMMs which are commercially available now. The techniques used by these sensors include LVDTs, optical gratings or interferometry, and have adopted essentially the same approach, i.e., the combination of three orthogonal high precision transducers to derive X, Y, Z coordinates. This approach requires only simple signal processing to achieve high accuracy. However they employ a complex mechanical design and are both bulky and expensive to manufacture. With the advent of microprocessors and microelectronic circuitry which provide for a high performance/cost ratio [2], a new mechatronic approach has been adopted. This involves reducing the mechanical complexity and a corresponding increase in signal processing stages to achieve the required accuracy. This results in a more compact unit (40mm diameter x 48mm length for the prototype) at lower cost without any compromise in ultimate performance.

2 OPERATING PRINCIPLE

2.1 The primary sensor

The primary sensor is shown in Figure 1. An array of seven optical fibres are mounted at the top through a termination in the cover, the position of which can be adjusted using precision side screws. In order to measure the 3D position, it is necessary to provide three degrees of freedom with the probe. There are a number of different methods available to achieve this [3]. The realization in Figure 1 is based on a polar coordinate system. Located on a linear ball bearing, the mirror cell is mounted at the upper end of the stylus, with a ball tip at the lower end of the stylus. A hemisphere is seated on three small ball bearings, which are fixed uniformly in the horizontal plane, providing the probe with three degrees of freedom (roll, pitch, and yaw). One more degree of freedom (linear translation) is provided by the

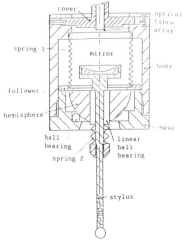

Fig 1   The primary sensor

linear ball bearing, which is fixed inside the hemisphere. Since only three degrees of freedom are needed, one of the four degrees of freedom, i.e., yaw rotation about the optical axis, should be removed by means of constraint. Two springs (spring 1 and spring 2) are employed to bring the stylus back to its rest position after the applied force is removed.

The above realization is a simple, easily fabricated, mechanical design. However, the flat surface of hemisphere and the top surface of the shoulder need to be aligned carefully by means of lapping the surfaces.

The effects of manufacturing tolerances as well as temperature expansion and wear on the probe performances can be analyzed as follows. They can be grouped into two categories, a) the change of supporting positions of the three ball bearings; b) three types of movement errors, i.e., play I between the follower and the flat surface of hemisphere, play II between the follower and the shoulder, play III between linear ball bearing and stylus shaft.

1) It can be shown that the horizontal positions of the three contacting points along a circle are not important. Assuming the three contacting positions $P1:(x_1,y_1,z_1)$, $P2:(x_2,y_2,z_2)$, $P3:(x_3,y_3,z_3)$, the pivot position (x_0,y_0,z_0) and the radius of hemisphere R, then:-

$$\begin{aligned}(x_1-x_0)^2+(y_1-y_0)^2+(z_1-z_0)^2=R^2\\(x_2-x_0)^2+(y_2-y_0)^2+(z_2-z_0)^2=R^2\\(x_3-x_0)^2+(y_3-y_0)^2+(z_3-z_0)^2=R^2\end{aligned} \quad (1)$$

It can be seen from Eq.(1) that the pivot position is defined by P1, P2, P3 and R. However, if $z_1=z_2=z_3=z$, i.e., in one horizontal plane, Eq. (1) becomes

$$\begin{aligned}(x_1-x_0)^2+(y_1-y_0)^2=r^2\\(x_2-x_0)^2+(y_2-y_0)^2=r^2\\(x_3-x_0)^2+(y_3-y_0)^2=r^2\end{aligned} \quad (2)$$

where $r^2=R^2-(z-z_0)^2$. This implies that as long as the three supporting positions are distributed around a horizontal circle, the pivot position will not be affected.

2) From Eq. (1), the effects of small variations of P1, P2, P3 and R on the variation of pivot point can be given as

$$\begin{bmatrix}\delta_{x0}\\\delta_{y0}\\\delta_{z0}\end{bmatrix}=\begin{bmatrix}x_{10}&y_{10}&z_{10}\\x_{20}&y_{20}&z_{20}\\x_{30}&y_{30}&z_{30}\end{bmatrix}^{-1}\begin{bmatrix}x_{10}\delta_{x1}+y_{10}\delta_{y1}+z_{10}\delta_{z1}-R\delta R\\x_{20}\delta_{x2}+y_{20}\delta_{y2}+z_{20}\delta_{z2}-R\delta R\\x_{30}\delta_{x3}+y_{30}\delta_{y3}+z_{30}\delta_{z3}-R\delta R\end{bmatrix}$$

$$------------------------------(3)$$

where $x_{i0}=(x_i-x_0)$, $y_{i0}=(y_i-y_0)$, $z_{i0}=(z_i-z_0)$, i=1,2,3; δX is the change of X, and X denotes x_j, y_j, z_j, and R, j=0,1,2,3.

3) The effects of play: The play I will cause an uncertainty of the rest position, which is permissable for 3D analogue probe because the stylus position can be directly measured. For triggering application, however, the rest position needs to be measured before touching the workpiece, thus allowing for software corrections. The play I also results in the possibility that the hemisphere may be lifted by spring 2 when the stylus moves upwards. This would lead to uncertainty of the mirror position and must be eliminated.

The play II will also cause the uncertainty of the rest position due to the spring (spring 1) hysteresis, but the hemisphere requires a larger force to produce lift than that required by the bearing shaft (By suitable choice of springs).

The play III will cause uncertainties of both rest position and other measurement positions. An interference fit is required between the stylus shaft and the linear ball bearing.

2.2 The modelling of the sensor

The light from an infra-red source is emitted towards the mirror by the central fibre, reflected by the mirror and detected by the remaining six optical fibres (Figure 1). The detected light intensities are modulated by the position and orientation of the mirror, which changes with the position of the stylus. The relationship between the light intensities and the stylus position is generally described by a system of non-linear equations.

The modelling of the sensor has previously been discussed [4], and the sensor model is described by the following formulae (4)-(7):

$$\begin{aligned}x_i&=-l\sin\theta\cos(\varphi);\\y_i&=-l\sin\theta\sin(\varphi);\quad(4)\\z_i&=l(1-\cos\theta)+\Delta z\end{aligned}$$

where l is the stylus length;
(x_i,y_i,z_i) is the position of stylus tip in the object coordinate system (OCS) defined in Figure 2;
$(\theta,\varphi,\Delta z)$ is defined in the sensor coordinates system (SCS), θ, the latitude, φ, the longitude, Δz, the displacement along z axis.

$$\begin{pmatrix}X\\Y\\Z\end{pmatrix}=\begin{pmatrix}l_1&m_1&n_1\\l_2&m_2&n_2\\l_3&m_3&n_3\end{pmatrix}(P_d-P_o)\quad(5)$$

where
(X,Y,Z) is the position of the detectors relative to the centre of the image plane. It varies with (x_i,y_i,z_i);
$(l_1,m_1,n_1),(l_2,m_2,n_2),(l_3,m_3,n_3)$ are the direction cosines of OX, OY, OZ in the sensor coordinate system (SCS), respectively. They are the functions of $(\theta,\varphi,\Delta z)$.
$P_d=(x,y,z)^T$ is the position of the detectors in SCS.
$P_o=(x_o,y_o,z_o)^T$ is the position of the image point in SCS. It is also a function of $(\theta,\varphi,\Delta z)$.

and

$$I(u,v)=(\frac{2}{u})^2[U_1^2(u,v)+U_2^2(u,v)]I_o\quad(6)$$

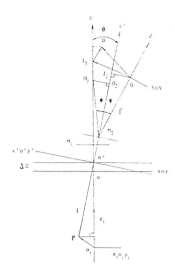

Fig 2   The relationship between the input position (x_i, y_i, z_i) and the detector's position relative to the centre of the image plane

or

$$I(u,v) = (\frac{2}{u})^2 [1 + V_o^2(u,v) + V_1^2(u,v)$$
$$-2V_o(u,v)\cos(\frac{u+\frac{v^2}{u}}{2}) - 2V_1(u,v)\sin(\frac{u+\frac{v^2}{u}}{2})$$
$$\hspace{2cm} (7)$$

where

$$u = \frac{2\pi}{\lambda}(\frac{a}{f})^2 Z$$
$$v = \frac{2\pi}{\lambda}(\frac{a}{f})\sqrt{X^2+Y^2}$$
$$U_n(u,v) = \sum_{s=0}^{\infty}(-1)^s (\frac{u}{v})^{n+2s} J_{n+2s}(v)$$
$$V_n(u,v) = \sum_{s=0}^{\infty}(-1)^s (\frac{v}{u})^{n+2s} J_{n+2s}(v)$$

and I is the intensity at the detector position, (X,Y,Z);
I_0 is the intensity at the centre of the image plane;
f is the distance from the mirror centre to the centre of the image plane;
λ is the light wavelength.

Given the input position (x_i, y_i, z_i), the light intensity detected by the fibres array $(I_1, I_2, I_3, I_4, I_5, I_6)$ (Figure 1) can be determined from the above sensor model.

An analysis of the model has predicted that a lateral resolution of 0.1 μm and depth resolution of 1 μm can be achieved assuming an SNR of 80 dB. An initial series of experiments were carried out to test the principles with an SNR of only about 60 dB. The results showed a lateral resolution of 0.5 μm and depth resolution of 5 μm [5]. This has shown good agreement between the modelling and the experimental results.

3   PROTOTYPE AND TEST

The schematic of the prototype sensor is shown in Figure 3. Optical feedback was used to stabilize the light source. In addition, a referencing channel provides for compensation of any residual light source drift. The light intensities detected by the optical fibres are amplified, filtered and sampled simultaneously, and buffered after a multiplexer stage. The A to D converter will then convert the signals to 14-bit digital signals channel by channel. The conversion time is less than 10 μs for each channel.

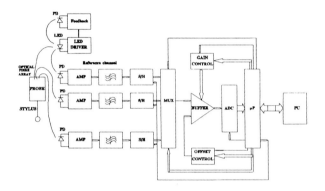

Fig 3   The schemitic of the sensor

Due to the manufacturing tolerances, the light intensities detected by the array varies significantly, but only the position of the entire optical fibre array can be adjusted. Therefore, an offset control must be provided to make the best use of the dynamic range of the analogue to digital converter (-3 to +3 v) for each channel. In addition, the measurement range of the sensor can be adjusted by means of an overall gain control. Finally, a single chip microprocessor is incorporated to analyze the digital signals and to provide for communication with a PC in addition the local intelligence.

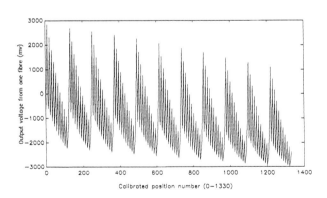

Fig 4   A typical signal detected by one of the optical fibres

Shown in Figure 4 is a typical signal obtained from one channel during a calibration, which has been carried out at 11x11x11 (1331) equally spaced positions within a cubic volume of range 200 x 200 x 200 μm. The calibration scans this cubic volume in a regular pattern of X, Y and Z sequences. Thus in Figure 3 each short duration transition corresponds to X axis displacement, the intermediate transition to Y axis displacement and the overall trend corresponding to Z axis displacement.

The noise (Figure 5) was also monitored during calibration. Clearly, the SNR of about 76 dB (i.e., 6000:1) has been achieved with the electronics.

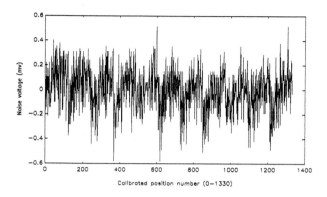

Fig 5   The noise in the sensor

The data obtained from the calibration is used to generate the empirical formulae using multivariate least square regression [6]. A 3-D non-linear error map is then produced to compensate for the sensor non-linearity. The algorithm for computing the measurement results is a hybrid approach based on the empirical formulae and an error map look-up table. The non-linear error is corrected by referring to the look-up table for each computed measurement obtained from the empirical formulae, on a reiterative basis. Ultimately the required position accuracy is achieved. The last computed result is accurate to the required precision and taken as the actual measurement. Normally sufficient accuracy is reached after 5 iterations; even in the worst case calculations are complete within 1 ms.

4 RESULTS

Following the calibration, ten boundary points and ten random points within the volume were chosen. Based on 50 samples for each point, X, Y, Z measurement repeatability (based on 2σ - 95% confidence limits) of better than 1μm has been uniformly achieved within the calibrated volume. Results of repeatabilities at some points indicates better performance than this and X, Y measurement repeatabilities are, as expected, somewhat better than the Z axis. The CMM has a limiting resolution of 1 micron which prevents determination of sub-micron performance.

Absolute accuracy (2σ) of each axis within the entire calibrated volume was 1μm on the prototype unit. The actual accuracy achieved is the same as the 1 μm repeatability limit.

5 CONCLUSION

The basic principle of precision 3-D displacement measurement by means of the controlled movement of a concave mirror in space has been established. Compact, readily manufacturable, sensors or probes for a wide variety of applications can be developed from this principle.

Due to its high precision, novel design, low cost and flexibility, the sensor will have many advantages over existing types, and exhibits great potential to solve many practical measurement problems in terms of accuracy, speed and flexibility.

Although primarily developed for coordinate measuring machines, high precision 3D position sensors are required for performance testing and in-cycle sensing on automatic assembly systems and robots. The probe will also provide advantages in terms of size and cost for machine tool applications.

REFERENCES

(1) Butler, C. An investigation into the performance of probes on coordinate measuring machines. Industrial Metrology, 1991, 2, 59-70.

(2) Giachino, J. M. Smart sensors. Sensors and Actuators, 1986, 10, 239-248.

(3) Petrucci, L. G. D. Metrology, 1986, 65-70. L & A Press, Solihull.

(4) Yang, Q. and Butler, C. Precision 3-D position measurement using a fibre optic sensor. Conference on Sensors and Their Applications (5th 1991 Edinburgh), Sensors: technology, systems and applications --(The Adam Hilger series on sensors), 1991, 281-286.

(5) Yang, Q. and Butler, C. Three-dimensional fibre-optic position sensor. Proceedings of International Conference on Optical Fibre Sensors in China OFS (C)'91, 1991, 558-563.

(6) Wang, L. et al Handbook of Mathematics, 1979, 851-852. Rimin Education Press, Beijing.

Intelligent transducers for materials physical properties measurement

G KULVIETIS, A DAUGELA, Vilnius Technical University, Lithuania

SYNOPSIS A computer-based mechatronic device for ultrasonic non-destructive testing applications is presented, using a piezoceramic converter suitable for dynamic measurements in the resonant state (in various forms and modes). The idea of the method is to evaluate the mechanical impedance of the material being tested with the aid of computer control.

NOTATION

$[K]$, $[C]$, $[M]$ - structural matrixes of stiffness, damping and masses;

$[T]$ - structural electromechanical matrix;

$[S]$ - structural matrix of capacity;

$\{F(t)\}$, $\{Pk\}$ - vectors of external electrical and contact forces;

$\{Q\}$ - vector of charges;

h, b, l - geometrical parameters of FE;

$\{I\}$ - vector of alternative current;

$\{Y\}$ - vector of conductivity;

H - indentation deepness (hardness);

R - indentor's radius;

E^* - complex Young's Modulus;

$n, \alpha, \beta, \{\delta\}, \{\varphi\}$ - approximation constants, vectors of displacements and potentials;

D - vector of electrical shrinkage;

e - tensor of dynamic coefficients of piezoceramic;

q - dielectric permeability of piezoceramic;

$1, 2, 3$ - local coordinate axis of the FE's node;

x, y, z - global coordinate axis of the FE's node;

i - number of the node.

1 INTRODUCTION

In numerous industrial applications it is important to register the distribution of the surface characteristics of machinery parts under the headings of hardness, rigidity and damping. The possibility of testing these had been suggested at least 20 years ago, but its implementation only became possible with recent advances in piezomechanics and signal processing by fast microprocessors. The mechanical state of structural materials and changes in structural elements in manufacturing operations can easily be described by two quantities of dynamic response: dynamic hardness and ductility of the material under contact conditions. Our attention has been concentrated on the problem of express non-destructive testing of polymers, elastomers and rubbers because these materials when heavily loaded can easily lose their properties with time (e.g. in exploitation and manufacturing). The main problem in testing polymers and rubbers is the fact that no measuring standards exist. The hardness of polymers and rubbers, which equates to the material's resistance to loading as well as the indentation depth of standard value contact loading, is the most informative parameter in express mechanical state testing. Young's Modulus and Loss factors could easily be calculated from the hardness value by using empirical expressions. We succeeded in building several transducer schemes applied for express dynamic nondestructive testing which are based on piezoactive converters. In the case of piezoactive converters, piezoceramic transducers performing double energy transformation (mechanic to electric and electric to mechanic) by using direct and inverse piezoeffects have been employed. The correlation between hardness, Young's modulus, loss factors and the piezotransducer's output signal have been established. During experiments it has been noticed that express and reliable analysis is available only in computer control due to the advance in software packages, which can easily evaluate several mechanical state characteristics by one indentation. It is possible to solve the executing model which includes theoretical and experimental results. The executing model is also involved with the problem of calibration. The model is based on simulation of the contact process in the zone of dynamic interaction in finite elements (FE).

2 PRINCIPAL OF THE METHOD

The schematic for surface measuring is shown in Fig.1. The mechanical measuring part of a system consists of a piezotransducer for testing and an actuator for applying load. In the case of a piezotransducer a piezoceramic-metal bimorphous converter suitable for bending oscillations has been used.

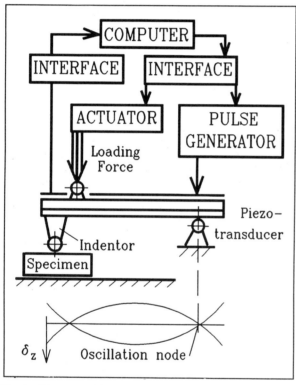

Fig 1   The measuring system

Usually resonant frequencies have been employed. The possibility for employing this transducer for polymers and rubbers testing can be easily investigated by using the basic theory of vibrations. The measuring process is dynamic interaction between the piezotransducer indentor and a specimen. The high frequency resonant state (usually a second or third mode, 2 . . . 20kHz) oscillations are generated by a sinusoidal voltage generator connected to the transducer input electrode. The output voltage is measured at the separated measuring electrode and due to the interface is evaluated by computer. Instead of the actuator for loading, a servo drive has been suggested. Computer control provides loading with various force value depending on the standard loading values. There are difficulties in comparing standards for hardness determination of polymers and rubbers. All comparisons have been based on the experimental curves and Scott equation approximations (this is application of Hertzian theory to the contact of polymers and rubbers).

The transducer, which works like the device mentioned above, has been used for measuring complex parameters - hardness and radial irregularities - in moving contact. The principal scheme of this converter shown in Fig. 2a. In this case a piezoceramic trimorphous transducer, which consist of two piezoceramics and a metal base suitable for continuous measurement in contact, has been proposed. The output signal, proportional to the complex parameter which consists of the geometric radial errors and material non homogeneity insertions equivalent to errors (see Fig. 2b.), has been evaluated by computer. The complex parameters of geometrical irregulations and insertions with various hardness for the moving roller could be simply described by using Fourier series. The transducer's high frequency resonant oscillations have been modelled by low frequency vibrations in a

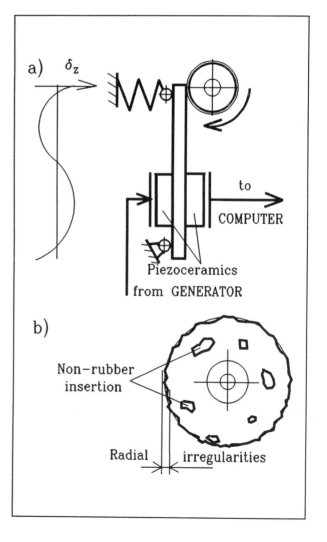

Fig 2   Continuous measuring transducer for rubber rollers testing

moving contact producing the output signal (it depends on angular velocity of the roller).

Both schemes provides quick and reliable testing.

3 MODELLING IN FE

The basic equation for FE modelling which includes piezoceramics influence is:

$$[K]\{\delta\}+[C]\{\dot{\delta}\}+[M]\{\ddot{\delta}\}-[T]^T\{\varphi\}=\{F(t)\}+\{P_k\}$$
$$[T]^T\{\varphi\}+[S]\{u\}=\{Q\}. \quad (1)$$

Following references (1), (2) and authors (3), (4) these matrix equations could be simply solved by using border conditions for the piezotransducer's electrode. The structural matrix of charges for the finite element shown in Fig. 3 could be written in the form:

$$\{Q\}=\int_0^l\int_{-h_0}^{h_1-h_0}D_3\,dz\,dx=$$
$$=-0.5e_{31}bh_1(h_1-2h_0)\times$$
$$\times\int_0^l\delta_{3x}dx+bh_1e_{31}\int_0^l\delta_{1x}dx+$$
$$+blq_{33}\{F(t)\}. \quad (2)$$

Structural matrixes of masses, stiffness and damping were built by using Ermit's functions (see (2)) of both the zero and the first levels. Contact force, which is based on the approximate data of the half-natural evaluation model, could be shown in the form:

$$P_{k_i}=nR^\alpha H^\beta E^*. \quad (3)$$

For example, Scott's experimental constants for the standard loading of rubber are equal to: n = 1.9, α = 0.65, β = 1.35. Contact compliances and their modelling during dynamic impact have been discussed in refs. (5), (6). Sinusoidally the alternating current effecting the piezotransducer's FE is:

$$\{I\}=jw\{Q\}. \quad (4)$$

Conductivity could be written as:

$$\{Y\}=jw(\{F\}^T\{\delta\}+blq_{13}). \quad (5)$$

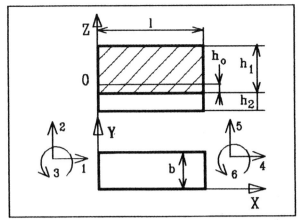

Fig 3 Finite element of the bimorphous (piezoceramic-metal) transducer

For the trimorphous piezoconvertor shown in Fig. 2 we can build Finite Element models as easily as for the bimorphous transducer.

4 RESULTS AND CONCLUSIONS

Some results of the investigation of the proposed measuring system have been presented in Figs. 4 and 5. Fig. 4 shows the calibration curves which have been based on the experimental data of the bimorphous piezotransducer. The experimental data are given in the form of ratios between output signal, load and hardness (in IRHD units). The approximated mesh of data (as mentioned above) has been employed to evaluate the main proportion between the loading force and indentation deepness by using the program. The indentor diameter for this transducer is 1 mm and the resonant frequency 4.12 kHz.

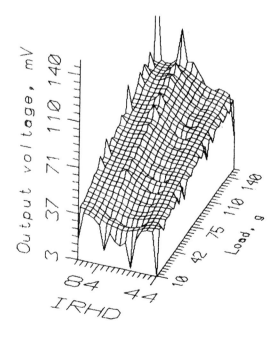

Fig 4 Calibration curves for the bimorphous piezotransducer

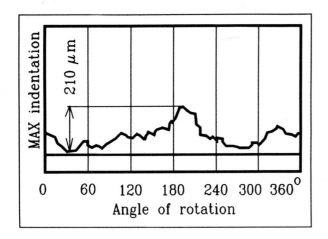

Fig 5 Typical results of the measuring identation depth for a rubber roller

Fig. 5 shows typical results of measuring the complex parameter (indentation depth) using a trimorphous piezotransducer applied for continuous testing. Indentor type - flat. Resonant frequency - 50.5 kHz. Diameter of the rubber roller is equal to 12 mm, angular velocity 0.4 rad/s.

The proposed method and program make it possible to perform rapid measurements of the mechanical state of the existing polymers and rubbers structures in exploitation and in the manufacturing process. The testing procedure consists of contacting the material through the dynamic impacts. By incorporating a dynamic model of a mechanical part of a system it is possible to evaluate variations in the mechanical behaviour inside the structures, something which cannot be easily achieved with the other nondestructive methods.

REFERENCES

(1) RAGULSKIS, K., BANSEVICIUS, R., BARAUSKAS, R., KULVIETIS, G. Vibromotors for Precision Microrobots. New York: Hemisphere Publishing, 1988.

(2) BARAUSKAS, R., KULVIETIS, G., RAGULSKIS, K. Vibromotors Design. Leningrad: Mashinostrojenije. 1984. (in russian).

(3) FILHO, F. V. Finite element analysis of structures under moving loads. Shock and Vibration Digest, 1978, 10, 27-35.

(4) KAGAWA, Y., YAMABUSHI, T. Finite Element Aproach to Energy - Trapped and Surface - Wawe Devices. IEEE Trans. on Sonics & Ultrasonics 1976, 4, 263-272.

(5) READ, E. R., DEAN, G. D., DUNCAN, J. C. Determination of Dynamic Moduli and Loss Factors. Elastic & Mechanical Properties. 1991, John Willey & Sons.

(6) VELUSHWAMI, M. CROSSLEY, F., HORVAY, G. Multiple Impacts of a Ball between two Plates. Trans.ASME,J.of Eng. for industry, 1975, 97, 828-835.

The use of giant magnetostrictive materials in fast-acting actuators

R D ADAMS, BSc, PhD, DSc(Eng), FInstP, FPRI, FIMechE, **C A McMAHON**, BSc, MIMechE, **R THOMAS**, BSc
University of Bristol, UK

SYNOPSIS

Research during the 1960's produced alloys exhibiting large magnetostrictive strains when subjected to a magnetic field. One example of this type of material is Terfenol, an alloy of iron and the rare earth elements terbium and dysprosium. The speed at which the material responds and the high force that it exerts makes Terfenol attractive for use in high-force, fast-acting actuators. This paper will characterise such actuators for mechatronic applications, and will review the relevant properties of Terfenol to identify the relationship between the driving parameters and the actuator displacement. The electromechanical and mechanical properties will also be considered to give some insight into the performance of these devices and into guide-lines for their design.

NOTATION

F_c critical frequency
ρ resistivity
d diameter of the rod
μ^S clamped permeability

1. INTRODUCTION

Giant magnetostrictive materials were developed as a result of research into alloys of rare earth elements and iron [1]. One such material, Terfenol (typically $Tb_xDy_{1-x}Fe_{1.95}$), has been found to have a maximum magnetostrictive strain of the order of 2×10^{-3}, when subjected to a magnetic field. This is approximately 30 times greater than the saturation magnetostrictive strain measured in annealed nickel.

The strain-field characteristics of Terfenol are dependent on the purity of the compound and the crystal structure. Two different compositions are most often available, $Tb_{.27}Dy_{.73}Fe_{1.95}$ and $Tb_{.3}Dy_{.7}Fe_{1.95}$. The latter gives a steeper initial increase in strain with applied magnetic field and a greater saturation strain but also exhibits more hysteresis.

The characteristics of Terfenol offer significant attractions as a material for actuators in mechatronic systems. Applications generally involve lengthwise straining of rods of the material whilst under compressive loading, to give a fast-acting, high-force but typically low-displacement actuator. Currently, one of the main applications for Terfenol is in sonar transducers [2][3], although many other applications are under investigation including precision positioning, sensors, servovalves and linear motors [4][5]. In this paper a classification of actuator characteristics will be presented, together with design guide-lines.

2. MAGNETOSTRICTION

Magnetostriction can be explained by the alignment of the magnetic domains within a material in the direction of an applied magnetic field. This movement causes a change in length of the material in the direction of the field. Since Terfenol increases in length, it is defined as positively magnetostrictive, whilst nickel, which contracts, is negatively magnetostrictive.

When Terfenol is put under compression, more domains align perpendicular to the direction of the load. Application of a magnetic field in the same direction as the load rotates more domains giving a larger output strain. If a field is applied in the opposite direction, the effect will still be an increase in length.

For a sinusoidal driving current, the material will strain at a fundamental frequency twice that of the frequency of the input current, as shown in Figure 1. To eliminate this frequency doubling, the Terfenol can be magnetically biased either by the use of a dc bias or with a permanent magnet.

3. PROPERTIES OF TERFENOL

Typical properties of Terfenol are given in Table 1. From the point of view of the actuator designer, a number of properties are particularly important:

o The coupling coefficient, which is a measure of the efficiency of conversion of electrical to mechanical energy, is high (in the order of 0.7), and is comparable with the most efficient piezo-electric ceramics.

o The relative permeability of the materials is, at 4.5(μ^S), very low compared to the value of 2000 for iron, so that high magnetic fields are needed to achieve the flux densities required for optimum operation. This permeability increases to 9.2 (μ^T) for a rigidly clamped material.

- Coupling between the electromechanical and mechanical properties also occurs in the value of Young's modulus. This effect is often known as the ΔE effect (6) and gives an increase of Young's modulus from 25GPa with no magnetic field to 70GPa.
- The mechanical damping of the material is high due to its magnetostrictive properties.

Table 1 Typical properties of Terfenol (7) (8)

Density	9.25×10^3 kg/m^3
Young's Modulus(Y^H)	25 GPa
(Y^B)	70 GPa
Curie Temperature	380°C
Coupling Factor	0.7
Relative permeability (μ_S^T)	$9.2 \times 4\pi \times 10^{-7}$ Tm/A
(μ^S)	$4.5 \times 4\pi \times 10^{-7}$ Tm/A
Resistivity(ρ)	60×10^{-8} ohm m
Energy Density	14000-25000 J/m^3

3.1 Published Results

Examples of the published magnetostrictive strain against field strength measured and reported by A.E.Clark (9) for $Tb_{.27}Dy_{.73}Fe_{1.95}$ are shown in Figure 2. The compressive load was achieved using a lever-arm type arrangement for loads up to 48.3MPa.

The increase in saturation strain with a rise in the applied magnetic field is shown clearly in these results. It can be seen that, as the pre-load increases, the magnetic field needed to obtain a given magnetostrictive strain increases also. For example, from Figure 2 (a) and (g), the field necessary to obtain a strain of 0.5×10^{-3} is approximately 20kA turns/m at 6.9MPa but this increases to close to 100kA turns/m at 48.3MPa.

4. ACTUATOR DESIGN

The main components of an actuator using a Terfenol rod as the active element are a driving coil and magnetic bias arrangement, a pre-load assembly, and a flux path. An example design is shown for a rig used for static and dynamic tests (Figure 3), which was designed to incorporate a Terfenol rod 6mm in diameter and approximately 75mm in length.

In designing an actuator, the following factors are important:

- Large driving coils are required to achieve the high magnetic fields (up to 300kA turns/m) required for saturation strain. Because the permeability is low, these should be wound closely to the rod, and should be made longer than the rod to reduce leakage effects.
- Permanent magnet bias is appropriate if the actuator is to be operated continually at or near the bias point, and if a constant value of bias is convenient. For variable or intermittent bias, a dc bias signal may be preferred.
- Mechanical loading may be provided by various means, for example by springs, dead-weights, loading rods or bars or hydraulic means. Springs and loading rods have the disadvantage that the compressive load is variable with deflection of the rod. In the test actuator, disc springs (Belville washers) were chosen since they are compact and a reasonably constant load could be achieved by careful consideration of the spring characteristics (10). However, disc springs may introduce hysteresis into the loading system.
- Ohmic heating may be such that cooling of the coil is required. In the test arrangement air cooling was used. For high power devices water cooled coils may be appropriate.
- The rods used in actuators should be machined with precision to ensure that any pre-load can be applied uniformly along the rod and to prevent any breakages in the rod. Special techniques are necessary for the machining of the rods due to the brittle nature of the material.

5. ACTUATOR PERFORMANCE

The experimental actuator was tested under conditions of static and dynamic applied field for a variety of pre-loads in order to establish whether the dynamic performance of the actuator could be predicted from knowledge of the static test results. Static tests involved rods of two different material compositions at different levels of pre-load and with applied magnetic fields up to 230kA turns/m. Dynamic performance was measured at various frequencies and with different pre-loads. The tests were completed first with no magnetic bias, the rod vibrating at twice the driving current, and then with an applied dc bias to eliminate the frequency doubling effect. In parallel, tests were conducted on an instrumented rod to investigate the temperature transients under varying electromagnetic loads.

5.1 Static Test Results

The static results for the rod with the composition $Tb_{.27}Dy_{.73}Fe_{1.95}$ (Figure 4) show that for an increased pre-load there is an increase in the maximum strain at 230kA turns/m. At 23.4MPa (Figure 4(c)), the relationship between the strain and the magnetic field is close to linear over a wider range of applied field than for lesser loads, allowing a greater range of the performance to be utilised, particularly when biased.

Results achieved with a Terfenol rod of composition $Tb_{.3}Dy_{.7}Fe_{1.95}$ (Figure 5) demonstrate that a greater maximum strain can be accomplished with this composition, although there is an increase in the hysteresis measured. In each case, it should be noted that although the strain is large for a magnetostrictive material, it is still small in absolute terms - perhaps 0.1mm displacement for a 75mm rod.

5.2 Dynamic Test Results

The performance of the actuator when operated at 100Hz and biased at approximately 25 kA turns/m is shown in Figure 6. The results showed that the maximum strain that can be achieved under dynamic conditions for a given applied magnetic field can be predicted from the static results. However, mechanical hysteresis was found to be significant in the experimental device. Design measures to reduce hysteresis are discussed in section 6 below.

5.3 Eddy Current Losses

The maximum frequency at which Terfenol can be operated efficiently is determined, generally, by the eddy current losses within the material. The critical frequency (F_c) at which these losses can become high is given by the expression

$$F_c = (2\rho)/(\pi\mu^s d^2)$$

For the 6mm diameter Terfenol rod, the critical frequency is 1.7kHz. In practice, Sewell and Kuhn (2) as well as Greenough and Schulze (11) have indicated that lamination of the Terfenol may be necessary above 1kHz. Sewell and Kuhn indicate that laminating the material is easily achieved, despite the brittle nature of Terfenol.

5.4 Temperature Measurement

The aim of the temperature tests was to measure the temperature distribution along the length of the rod under typical operating conditions. Six thermocouples were placed along the length of a rod, which was excited by a similar coil to that being used in the experimental rig. The magnetic field was applied for short bursts of 5 or 10 seconds duration every minute and the temperatures were recorded at each of the six thermocouples 30 seconds after each excitation. After one hour, there was no further excitation and the temperature drop was recorded.

Driving the Terfenol rod at 1Hz and 13.5kA turns/m produced no significant rise in temperature. However, at 150Hz and 500Hz (Figure 7) there was a significant rise in temperature initially, followed by a gradual increase over the period of the test. Once the excitation had stopped the cooling was rapid. The rapid rise in the temperature of the rod following the excitation and the lower increase in temperature for lower frequencies as seen in Figure 7(b), suggest that the heating of the Terfenol is associated with eddy current losses within the material. The temperature climbs towards a maximum for each frequency, as the rate of cooling approaches the rate of heating.

J.L.Butler (7) has reported a decrease in the magnetostrictive strain with temperature for the composition $Tb_{.3}Dy_{.7}Fe_{1.95}$, indicating that temperature may be a factor in the performance of an actuator with Terfenol as the active element. In his investigations, Butler found that, with an applied magnetic field of approximately 160 kA turns/m and a pre-load of 7.6MPa, there was a reduction in magnetostrictive strain from close to 1.7×10^{-3} at 0°C to about 1.3×10^{-3} at 70°C. The change in strain was less for lower fields, indicating that the effect is to reduce the maximum strain that can be achieved. The decrease in strain was found to be greater with a higher compressive load.

6. INCORPORATION INTO MECHATRONIC SYSTEMS

It has been shown that actuators using Terfenol have the characteristic of high force over small displacements at frequencies without lamination of up to about 1500Hz, and with lamination very much higher than this. If appropriately biased, the strain/field characteristics of the material are near linear under certain loading conditions. Very steep strain/field curves can be achieved for certain materials by careful consideration of these loading conditions.

These characteristics have allowed the development of a range of actuation devices with a variety of applications that may be broadly categorised as follows:

o **High force vibratory or positioning devices**: Terfenol-based actuators offer very much higher forces than equivalent moving coil devices, and offer larger displacement and therefore higher power than piezo-ceramic materials, in particular at the lower end of the frequency spectrum (below 3000 Hz).

o **Switching devices**: If biased close to the steepest part of the strain/field curve, Terfenol offers very high switching rates with operation times in the order of 1ms and with high loads. For example, with suitable mechanical amplification, Terfenol can provide displacements equivalent to those of rotary solenoids at 10-50 times the rate.

o **Inch-worm devices**: By using electromagnetic inch-worm mechanisms (4), high force actuators with minimum moving parts and potentially high positioning accuracy may be constructed.

The critical design features for a successful actuator are as follows: careful choice of mechanical and electrical bias points; the use of solid pre-loading devices (eg loading rods), and the minimum number of joints in order to minimise hysteresis; careful magnetic design (almost certainly with the assistance of Finite Element modelling) to account for the low permeability of Terfenol, and lamination of the rod if necessary. Even with careful design, however, there are a number of constraints:

o The cost of the material is high;
o The high magnetic fields that are required mean that large, highly inductive coils are necessary. This imposes a significant constraint on the power amplifier and controller design, which is further complicated by the non-linear displacement/field characteristic of the material.
o For many applications a mechanical amplification system is required. Very careful mechanical design of the amplification system is required. The dynamic characteristics of the actuator and the effect of the dynamic mechanical loads on the electromagnetic performance of the actuator should be considered as part of this design task.

7. CONCLUSIONS

The rare earth alloy Terfenol may be used to construct electromagnetic actuators which offer faster actuation times than moving iron actuators, higher actuation forces than moving coil devices, and larger displacements than are possible with piezo-ceramic materials. This paper has discussed the strain/field characteristics of the material, together with aspects of the magnetic and mechanical design of actuators. In particular, the importance of design for the low electromagnetic permeability of the material, and the importance of eddy current, temperature and control considerations, has been noted.

ACKNOWLEDGEMENTS

The work reported in this paper has been supported by the Science and Engineering Research Council as part of grant GR/G 59004.

REFERENCES

(1) Clarke, A.E. Ferromagnetic Materials, 1980, Vol.1, chapter 7, pp 531-587 (North-Holland Publishing Company).

(2) Sewell, J.M. and Kuhn, P.M. Opportunities and Challenges in the Use of Terfenol for Sonar Transducers, Power Sonic and Transducer Design Workshop, Lille, 1987, pp 134-142.

(3) Boucher, D. Trends and Problems in Low Frequency Sonar Projector Designs, Power and Transducer Design Workshop, Lille, 1987, pp 100-119.

(4) Fahlander, M. and Richardson, M. New Material for the Rapid Conversion of Electric Energy to Mechanical Motion, 1989 (Feredyn AB).

(5) Dyberg, J. Magnetostrictive Rods in Mechanical Applications, First International Conference on Giant Magnetostrictive Alloys and Their Impact on Actuator and Sensor Technology, Marbella, 1986, pp 193-214.

(6) Bozorth, R.M., Ferromagnetism, 1951, pp 684-699, (Van Nostrand Company).

(7) Butler, J.L. Application Manual for the Design of Etrema Terfenol-D$_{TM}$ Magnetostrictive Transducers, 1988 (Edge Technologies).

(8) Johnson-Matthey Datasheet number 8.

(9) Clark, A.E., Spano, M.L. and Savage, H.T. Effect of Stress on the Magnetostriction and Magnetisation of Rare Earth-Fe$_{1.95}$ Alloys, IEEE Transactions on Magnetics, 1983, Vol. Mag-19, No.5.

(10) Brown, A.A.D. Mechanical Springs (Engineering Design Guides;42), 1981, pp 26-31 (Oxford University Press).

(11) Greenough, R.D. and Schulze, M. AC Losses in Grain Orientated Tb$_{0.3}$Dy$_{0.7}$Fe$_{1.95}$, Department of Applied Physics, University of Hull.

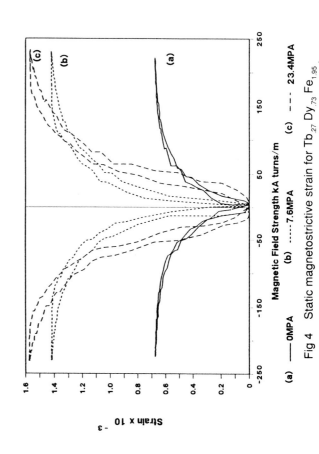

Fig 1   Magnetostrictive strain versus magnetic field strength, demonstrating the frequency doubling effect

Fig 2   Magnetostrictive strain versus applied magnetic field for Tb$_{.27}$Dy$_{.73}$FE$_{1.95}$, published by A.E. Clark (6)

(a) 6.9 MPa
(b) 13.8 MPa
(c) 20.7 MPa
(d) 27.6 MPa
(e) 34.5 MPa
(f) 41.4 MPa
(g) 48.3 MPa

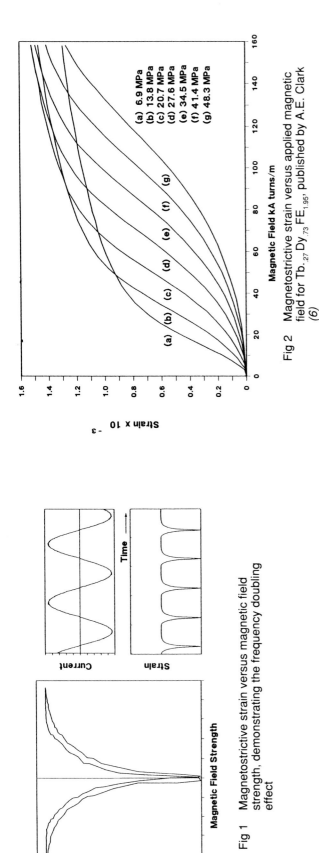

Fig 3   Magnetostriction test rig

Fig 4   Static magnetostrictive strain for Tb$_{.27}$Dy$_{.73}$Fe$_{1.95}$

(a) —— 0MPA   (b) ----- 7.6MPA   (c) --- 23.4MPA

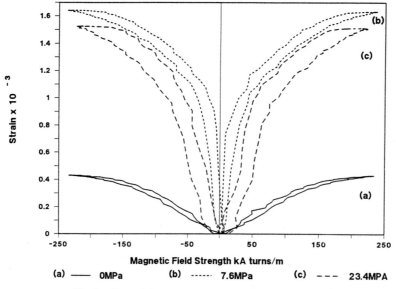

Fig 5 Static Magnetostrictive Strain For $Tb_{.3}Dy_{.7}Fe_{1.95}$

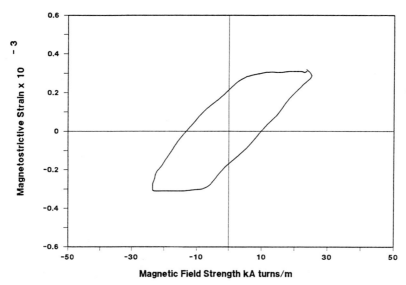

Fig 6 Dynamic magnetostrictive strain versus applied magnetic field for $Tb_{.27}Dy_{.73}Fe_{1.95}$, pre-loaded to 7.6MPa, biased at approximately 25 kA turns/m and operated at 100Hz

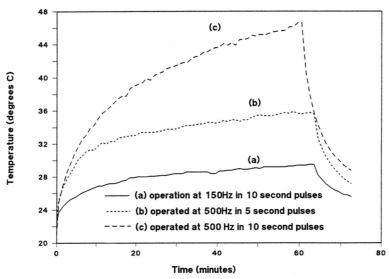

Fig 7 Temperature behaviour of Terfenol within an actuator

Flexible joint control of a KUKA IR 161/60 industrial robot

J SWEVERS, D TORFS, M ADAMS, J DE SCHUTTER, H VAN BRUSSEL
Katholieke Universiteit Leuven, Belgium

SYNOPSIS   This paper presents two modifications of the standard robot controller for a KUKA IR 161/60 industrial robot : improved trajectory generation and flexible control of the first three joints. The flexible controller for the first joint is a robot configuration dependent combination of linear state feedback controllers with feedforward. The flexible controller for the second and third axis is a nonlinear decoupling controller. Tests show that the improved trajectory generation gives the largest contribution to the improvement of the performance, and that only at very high velocities and accelerations, there is a significant difference between a flexible controller and a classical PID controller.

1 INTRODUCTION

During the last decade, much attention has been paid to flexible robot control. A lot of control algorithms and approaches have been put forward. Most of them are tested by simulation or on idealized test setups. These test cases are mostly "very" flexible, light weight robots, intended for space applications.

The work described in this paper is carried out as part of **Esprit**-project **1561** ("SACODY"), that is concerned with industrial robotics. This paper focuses on flexible robot controllers for an **industrial robot**, more precisely a KUKA IR 161/60 [1].

Industrial robots have specific characteristics, limitations, and fields of application that differ from those of the flexible robots considered by most researchers.

- The construction of industrial robots aims at maximum stiffness. Their flexibility is very limited. The expected improvement in dynamical performance, obtained by changing from a rigid robot controller to a flexible robot controller, is small.

- Industrial robots are far from idealized structures. Coulomb friction, stick-slip, backlash, not clearly defined and robot-orientation-dependent mode shapes and resonance frequencies are some robot characteristics that hamper the modelling. Moreover, flexible systems are distributed parameter systems and therefore contain an infinite number of modes. Only a few of these modes can be included in the modelling and control design. Flexible robot controllers that require a complete and accurate model have to be avoided.

- Implementation on a present-generation industrial robot control unit has to be considered if the improved robot controllers are to be used in industrial environments. The present-generation industrial robot control units are not very powerful. Their computational power is not comparable to the computer systems used in laboratories. This imposes severe restrictions on the complexity of the control algorithm.

- An industrial environment and application limits the types of noncollocated sensors that can be used. Direct end effector position measurement, for example by means of a 3D laser tracking system [2] or a camera vision system [3], is feasible only in laboratory environments or space applications.

Accelerometers, attached to the end effector or to the different links, are not very reliable in industrial environments, because of their noise sensitivity and the strong electro-magnetic disturbances often present in industrial environments. Therefore, accelerometer signals have to be filtered before they can be used in feedback control [4].

For a wide class of industrial robots, experimental results reveal that joint flexibility rather than link compliance is the dominating source of robot flexibility [5,6,7,8]. Strain gauges or real-time optical link deflection measurement systems, such as **DIOMEDES** [9,10], cannot be used. A reliable and accurate joint deformations measuring technique is to use an extra encoder attached to the link in addition to the encoder mounted on the motor shaft [5]. The difference between the motor encoder signal and the corresponding link encoder signal is then a measure of the deformation of the joint. The latter technique is applied in this paper to control the flexibility of the first three joints of a KUKA IR 161/60 industrial robot.

- In industrial robot applications, continuous path control has become more and more important. Continuous path control requires accurate tracking of a predetermined continuous path. It is used for applications such as arc welding, grinding, painting, and pasting. Point-to-point control requires only positioning accuracy. It is used for applications such as parts handling and spot welding. This difference has a consequence for the controller. Feedback can accomplish good positioning accuracy and is sufficient for point-to-point applications. It is not sufficient to guarantee fast and accurate tracking, since high feedback gains do not eliminate tracking errors. Feedforward has to be included to this end.

This paper describes two modifications of the standard controller for a KUKA IR 161/60 industrial robot : improved trajectory generation and flexible control of the first three joints. The modelling and control design consider only the flexibility of the first three joints because these joints exhibit the main robot flexibility. The flexibility of the other joints and the link flexibilities are neglected. Joint flexibility is modelled as a torsional spring and damper, so only one flexible mode per joint is taken into account. The weak dynamic coupling between the first axis (vertical axis) and the remaining axes is neglected.

The flexible controller for the first joint is a robot configuration dependent combination of linear state feedback controllers with feedforward. It is simple and can be implemented on an industrial robot control unit. The flexible controller for the second and third axis is a nonlinear decoupling controller. It is more complex such that its implementation requires a more powerful control unit.

Tests, which are based on the test standards of ISO (Industrial Organisation for Standardization), and RIA (Robotic Industries Association), show that the improved trajectory generation gives the largest contribution to the improvement of the performance. Only for tasks that require very high velocities and accelerations, there is a difference between the performance of a flexible robot controller and a PID controller.

Section 2 describes the KUKA robot, the sensor equipment and control hardware. Section 3 discusses the different robot controllers that have been tested. Section 4 reports and interprets the test results.

2 DESCRIPTION OF THE TEST SETUP

Fig 1 shows the setup for one of the tests. The robot is a KUKA IR 161/60 industrial robot (with an arm extension AV400): it has 6 axes, it has a working envelope with radius 3150 mm, and it is equipped with a spot welding tool of 45 kg. Three extra encoders are attached to the first three links of the robot. They measure the link position in addition to the motor position, which is measured with an encoder mounted on the motor shaft. The difference between the motor encoder signal and the corresponding extra encoder signal is a measure for the deformation of the joint. An experimental modal analysis of this robot has shown that the main flexibility in this robot is caused by the first three joints [6], such that these extra encoders suffice to measure the flexibility. A digitizing pen and tablet together with the software package RODYM [11]

Fig 1 KUKA IR 161/60 industrial robot equiped with spot welding tool

measure the end effector movements and positioning. The digitizing pen is attached to the tool of the robot.

The improved trajectory generation and the control algorithms are programmed on a VME-system. This system can read encoder signals and send out analog voltages in the range of $\pm 10\,volt$. These voltages are linearly converted by the power supply of the motors into a motor current, when the power supply is working in torque command, or a motor velocity, when the power supply is working in velocity command. The command mode can be altered for each axis separately.

3 ROBOT CONTROL

A robot controller consists of two levels: a system level and a servo level. From a specified task (e.g. move from point A to point B on a straight line) the system level generates trajectories for all the joint angles. Two different trajectory generation algorithms have been implemented. The first algorithm is based on a constant acceleration and deceleration profile, and is the standard trajectory generation algorithm for an industrial robot. The second trajectory generation algorithm is based on a 9th order polynomial. It results in a smoother trajectory for the different joint angles.

The output of the system level, i.e. the trajectories for all joint angles, is the input for the servo level. The servo level controls the different motors according to specified algorithms such that the joint angles follow the desired joint angles.

Four different robot controllers with increasing level of complexity are tested.

(1) Standard robot controller

The first robot controller is the standard robot controller which is normally delivered with the robot. It controls all axes in velocity command with independent PID controllers with velocity feedforward, and uses a trajectory based on a trapezoidal velocity profile. This controller is also implemented on the VME-system.

(2) Standard robot controller with improved trajectory

The second controller has the same servo level as the first one, but uses a trajectory based on a 9th order polynomial.

(3) Robot configuration dependent state feedback controller for axis 1

The third robot controller controls axis 1 in torque command using state feedback and feedforward, axes 2, 3, 4, 5, and 6 in velocity command with independent PID controllers with velocity feedforward, and uses a trajectory based on a 9th order polynomial. The state feedback controller is based on 4 fourth order linear models of the robot. Each of these models have been identified experimentally for a certain configuration of axes 2 and 3. These configurations lie between the fully extended configuration and fully contracted configuration of the robot. The used identification method is the weighted least squares frequency domain identification method described in [12]. The identifications are based on stepped sine measured frequency response functions.

The models take into account the flexibility of the first joint with one resonance and one anti-resonance frequency. The identification of each fourth order state space model is based on the identification of two discrete time transfer functions. The first transfer function relates the input command to the deformation of the flexibility, i.e. the difference between the angular link and motor position of the first axis. This transfer function contains a pair of complex conjugated poles that describes the resonance frequency. The second transfer function relates the difference between the angular link and motor position of the first axis and the angular position of the first axis. This transfer function contains a double integration (two poles at $z = 1$) and a pair of complex conjugated zeros that describes the anti-resonance frequency. The division of the total model into these two submodels and their separate identification results in a more accurate total model than the direct identification of the total model based on one frequency response function [13].

The control signal $u[k]$ for the first axis is a weighted sum of four control signals $u_j[k]$:

$$u[k] = \sum_{j=1}^{4} w_j[k] u_j[k]. \qquad (1)$$

$w_j[k]$ is the weight for $u_j[k]$. The control signals $u_j[k]$ consist of feedback of the difference between the desired and measured states, and feedforward:

$$u_j[k] = u_{ff j}[k] - \mathbf{K}_j(\mathbf{x}[k] - \mathbf{x}_{d\,j}[k]). \qquad (2)$$

The design of the state feedback controllers and the calculation of the feedforward signals are based on the identified state space models. All state variables are measurable. The feedback gains are calculated using pole placement. The calculation of the feedforward signals is a simulation of the inverse state space models with the desired trajectory. The desired state trajectories $\mathbf{x}_{dj}[k]$ result from a simulation of the state space models using the calculated feedforward signals.

The weights $w_j[k]$ provide a smooth switching between the different state feedback controllers. They are function of the trajectories for axes 2 and 3. The measure used in this function is an estimation of the distance $d[k]$ of the end effector from the first axis of rotation. The calculation of this distance does not take into account wrist axes rotations, for simplicity.

(4) Nonlinear decoupling controller for axes 2 and 3

The fourth robot controller is used in one test only. It only controls axes 2 and 3 in torque mode with a nonlinear control algorithm that takes the flexibility of these joints into account [14]. The control algorithm is based on the full nonlinear model of links 2 and 3 of the robot. This model is derived applying Lagrange's equations. The elastic coupling between the links and the joints is modelled as a torsional spring. The motor friction is modelled as a sum of linear viscous friction and Coulomb friction. The identification of the parameters in this model is performed in two steps. (1) First the parameters related to the rigid body model are estimated. (2) Then the spring constants are estimated by using static, quasi-static, noise and stepped sine measurements.

The applied nonlinear decoupling controller [14] consists of three stages: (1) a nonlinear feedforward to cancel out nonlinear terms in the rigid body dynamics, (2) a linear feedback of all state variables, i.e. rigid and flexible coordinates, (3) a compensation of the static and dynamic deflection by calculation of the reference values of all state variables based on the model of the flexible robot. The controller uses besides the motor encoders also the extra encoders of the these joints. The trajectory is based on a 9th order polynomial.

4 DESCRIPTION OF THE TESTS

Three sets of tests have been executed to evaluate the different robot controllers: (1) Pose Stabilisation Time and Pose Overshoot; (2) Minimum Positioning Time (time required to travel between two points); (3) Test contour for industrial robots: a circle and a straight line. These tests are described in the test standards of ISO (Industrial Organisation for Standardization) [15] and RIA (Robotic Industries Association) [16] or are internal KUKA tests [17].

The following sections give a brief description of the different tests.

4.1 Pose stabilization time and pose overshoot

The **pose stabilization time** is, according to the ISO 9283 standard [15], the period of time which elapses between the instant at which the robot end effector reaches the desired position for the first time and the instant at which the damped oscillatory motion of the end effector falls within a specified limit around that specified position.

This limit is called the repeatability zone. According to the same standard, the **pose overshoot** is the maximum value of the absolute distance between the end effector positions reached after crossing the final position for the first time and the final end effector position. The RODYM [11] system measures the settling time instead of the pose stabilization time, and the pose overshoot. The **settling time** is, according to the RIA standard [16], the period of time which elapses between the instant at which the end effector enters a specified limit band around the desired position for the first time and the time instant of staying inside this band. There is little difference between the settling time and the pose stabilization time in our tests, such that they can be considered equivalent.

The ISO standard 9283 defines five positions, located in a plane of a cube in the working space of the robot, between which these tests have to be performed. All pose characteristics have to be tested at the maximum velocity achievable between the specified poses. Only two of the 10 possible tests are reported in this paper. These two tests are the most critical of all pose stabilization tests because the points of arrival are close to the boundary of the working space.

An extra pose stabilization test has been added to these tests. This test evaluates the dynamic performance of the different controllers for the first axis. This because the first joint is the most flexible part of the robot. The robot is put in its fully extended configuration for this test, such that the moment of inertia with respect to the first axis is maximum. This results in a minimal eigenfrequency and a maximum end point oscillation. In this test, the first axis moves at maximum achievable velocity over 90 degrees while all other axes have their brakes on.

4.2 Minimum positioning time

The minimum positioning time is the time between departure from and arrival at a certain position after covering a predetermined distance (from 10 mm to 500 mm) under point to point (PTP) control. The RODYM measurement system calculates the minimum positioning time as the time between leaving and entering a circle around start point and destination point. It is clear that the difference between the minimum positioning time according to the ISO standards and the time measured by the RODYM system is small if the radius of this circle is small in comparison to the covered distance.

The measurements are performed under continuous path (CP) control instead of PTP control to avoid touching or even damaging the digitizing tablet.

4.3 Test contour for industrial robots

The test contour is a circle with radius 80 mm that lies in a horizontal plane. It is a critical part of a complex test contour for industrial robots, specified in [17], This part of the test contour is sufficient to evaluate the path characteristics of robot controllers. It is executed at low velocity, i.e. 10% of the maximum velocity, and at maximum velocity. The end effector path measured at low velocity is considered as the reference path, with which the maximum velocity path is compared.

An extra tracking test has been added to these tests to evaluate the nonlinear controller for axes 2 and 3: a straight line of 225 mm. The line makes an angle of 45 degrees with the horizontal plane. In this test only axis 2 and 3 are controlled, while axes 1, 4, 5, and 6 have their brakes on.

Table 1 Results of the pose stablization tests

controller	TEST 1	
	settling time (s)	overshoot (mm)
controller 1	0.557	2.417
controller 2	0.000	0.280
controller 3	0.000	0.280
	TEST 2	
	settling time (s)	overshoot (mm)
controller 1	0.300	0.792
controller 2	0.000	0.275
controller 3	0.000	0.000

5 CONTROL RESULTS

This section discusses the results of the tests described in the previous section.

5.1 Pose stabilization time test results

Table 1 gives the results of the pose stabilization time and pose overshoot tests. Each measurement is repeated three times and the mean values are reported. The value of the repeatability zone is 0.3 mm. Fig 2, 3, and 4 show the results of the extra pose stabilization test. For controller 1, there is an overshoot of 1.60 mm and the settling time is 0.65 seconds. For controller 2, there is an overshoot of 0.56 mm and the settling time is 0.16 seconds. For controller 3, there is no overshoot and the end effector does not leave the repeatability zone. This corresponds to a settling time of 0.0 sec.

These results (table 1) lead to the conclusion that a smooth trajectory gives the most important contribution to an improved dynamic behaviour (compare controllers 1 and 2). Only at extreme high velocities, such as in the extra test (Fig 2, 3, and 3), is there a significant difference between controller 2 and 3. In that case there is still an excitation of the structural dynamics if the controller does not take the flexibility into account.

5.2 Minimum positioning time

The average minimum positioning time for controllers 2 and 3 is 22 % smaller than for controller 1. These longer positioning times for controller 1 are necessary to avoid excitation of the structural dynamics.

5.3 Test contour for industrial robots

The two velocity levels mentioned in this section correspond to 10 % and 100 % of the maximum velocity allowed

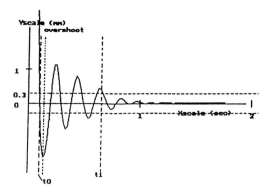

Fig 2   End effector response measured during extra pose stablization test with controller 1

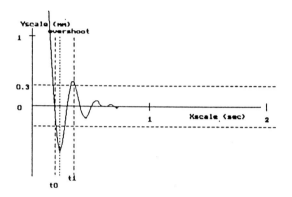

Fig 3   End effector response measured during extra pose stabilization test with controller 2

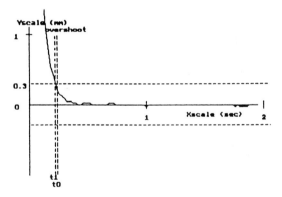

Fig 4   End effector response measured during extra pose stabilization test with controller 3

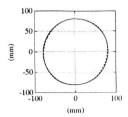 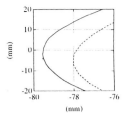

Fig 5   End effector position measured during circle tracking test with controller 1; low velocity: solid line, high velocity: dashed line

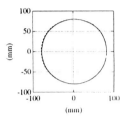 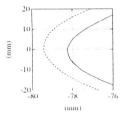

Fig 6   End effector position measured during circle tracking test with controller 2; low velocity: solid line, high velocity: dashed line

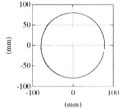 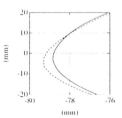

Fig 7   End effector position measured during circle tracking test with controller 3; low velocity: solid line, high velocity: dashed line

by the standard controller.

Fig 5, 6, and 7 show the circle tracking test results of the first three controllers. The right figure is a detail of the left figure, taken at the point where the velocity reaches its maximum. The solid line is the result obtained at low velocity, and the dashed line is the result obtained at high velocity. The maximum deviation for controller 1 is 1.6 mm, for controller 2 1.1 mm, and for controller 3 0.4 mm.

Fig 8 shows the straight line tracking test results of controller 2 and 4. The static error due to the gravitation is compensated for controller 4. Fig 9 compares the dynamic tip position errors normal and tangential to the desired trajectory for controller 2 and 4. The maximum dynamic tracking error normal to the desired trajectory is reduced from 1.6 mm to 0.8 mm. The maximum dynamic tracking error tangential to the desired trajectory is reduced from 2.5 mm to 2.0 mm. The normal and tangential static errors due to gravitation are compensated.

6   CONCLUSION

The tests have shown that the performance of industrial robots can be improved with simple modifications of the industrial robot controller. The change from a trajectory based on a trapezoidal velocity profile to a smooth trajectory gives the largest contribution to the improved dynamic performance. Only at very high velocities and accelerations, is there a significant improvement of the flexible controller over a classical PID controller in terms of overshoot and oscillations.

The tests have also shown that a nonlinear flexible controller for axes 2 and 3 compensates the static deflections. The improvement of the dynamic accuracy is however small, due to the limited flexibility of joints 2 and 3 and the low velocity and acceleration during the tests.

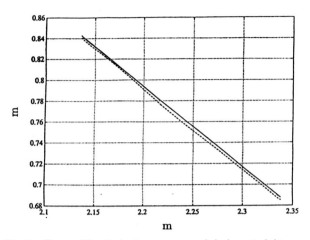

Fig 8 Tip position trajectory measured during straight line tracking test using controller 2 (dashed line) and controller 4 (dotted line), compared with the desired trajectory (solid line)

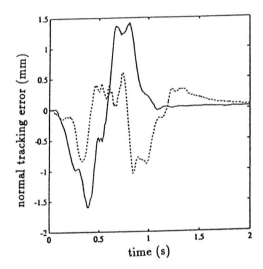

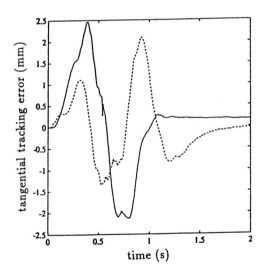

Fig 9 Tracking error normal and tangential to the desired trajectory in function of time for controller 2 (solid line) and controller 4 (dashed line)

REFERENCES

(1) KUKA SCHWEISSANLAGEN UND ROBOTER GmbH, KUKA Industrieroboter Typ IR 161/60. Robot specification, Dok 065.05.4d.

(2) VAN BRUSSEL, H. Identification and control of flexible structures. Proc. Int. Conf. on Advanced Mechatronics, JSPE, Tokyo, 1989, pp. 347-352.

(3) CANNON, R.H., SCHMITZ E. Initial experiments on the end-point control of a flexible one link robot. Int. J. of Robotics Research, 1984, Vol. 3, No. 3, pp. 62-77.

(4) FUTAMI, S., KYURA, N., HARA, S. Vibration absorption control of industrial robots by acceleration feedback. IEEE Transactions on Industrial Electronics, 1983, Vol. IE-30, No. 3, pp. 299-305.

(5) KUNTZE, H.B., JACUBASCH, A.H. Control algorithms for stiffening an elastic industrial robot. IEEE J. of Robotics and Automation, 1985, Vol. RA-1, No. 2, pp. 71-78.

(6) VAN DER AUWERAER, H. Off-line identification of a KUKA IR 161/60 industrial robot. 1990, Technical Report, task 4.8.5.1., Esprit project 1561: Sacody.

(7) WIDMANN, G.R., AHMAD, S. Control of industrial robots with flexible joints. Proc. IEEE Conf. on Robotics and Automation, Raleigh, North Carolina, March 31 - April 3 1987, pp. 1561-1566.

(8) RIVIN, E.I. Effective rigidity of robot structures: analysis and enhancement. Proc. of the 1985 American Control Conference, Boston, June 1985, pp. 381-382.

(9) DEMEESTER, F., VAN BRUSSEL, H. Real-time optical measurements of robot structural deflections. Mechatronics, 1991, Vol. 1, No. 1, pp. 73-86.

(10) SWEVERS, J., TORFS, D., DEMEESTER, F., VAN BRUSSEL, H. Fast and accurate tracking control of a flexible one-link robot based on real-time link deflection measurements. Mechatronics, 1991, Vol. 2, No. 1, pp. 29-41.

(11) VAN DEN BOSSCHE, J. RODYM: a new approach to robot metrology. 1989, internal report 90P33, K.U. Leuven, Department of Mechanical Engineering.

(12) SWEVERS, J. Linear identification and control of flexible robots. Ph.D. Thesis, Januari 1992, Department of Mechanical Engineering, K.U.Leuven, Belgium.

(13) SWEVERS, J., TORFS, D., DE SCHUTTER, J., VAN BRUSSEL, H. Accurate tracking control of a flexible one-link robot. Proc. of the 15th In-

ternational Seminar on Modal Analysis, Leuven, Belgium, 19-21 September 1990, pp. 353-366.

(14) DE SCHUTTER, J., VAN BRUSSEL, H., ADAMS, M., FROMENT, A., FAILLOT, J.L. Control of flexible robots using generalized non-linear decoupling. Proc. Symposium on Robot control, Karlsruhe, 1988, pp. 98.1-98.6.

(15) ISO/9283: MANIPULATING INDUSTRIAL ROBOTS Performance criteria and related testing methods.

(16) ANSI/RIA R15.05 Evaluation of the static performance characteristics for the comparison of robots and robotic systems, Revision 10.

(17) GERUNG, M. Accuracy measurement on a KUKA IR 161/60 with RCM3. 1990, Technical Report, Esprit project 1561: Sacody.

Integrated process and control design for fast coordinate measuring machine

M R KATEBI, MSc, PhD, MIEE, CEng, **J McINTYRE**, BSc, **T LEE**, **M J GRIMBLE**, PhD
Strathclyde University, UK.

ABSTRACT

Coordinate Measuring Machines (CMM) are widely used in manufacturing processes for accurate and rapid inspection of machine products. Accuracy and speed are two conflicting requirements of the CMM control system which are dictated by the servo control loops and dynamics of the arms. The CMM considered here consists of three independent servo controlled orthogonal arms, a two Degree-Of-Freedom (DOF) probe, a granite table and a computer for man-machine interface. The objective of this paper is to develop an optimal control solution to accurately position the probe at minimum time.

Each servo loop is constructed of a pre-amplifier, a Pulse Width Modulated (PWM) current amplifier and a three-phase brushless DC motor. The pre-amplifier houses the velocity loop control electronics. Velocity is detected through an integral optical shaft encoder. A 20 lines/mm optical grating is used for position measurement. A sensor is developed to give 0.05 mm positioning accuracy. The interface between the system and the controller is through a commercially available input/output card programmed through high level software.

The nonlinear dynamics of each components of the system are modelled and simulated to gain better understanding of the system and to set up a design environment for the machine and control system. This model is subsequently linearised and employed to design optimal position controller for the servo system.

The control design is restricted to simple PID and Lead-Lag controller in order to keep the computation time to minimum. Simulation and experimental data are presented to demonstrate the proposed control design procedure.

1 Introduction

Coordinate Measuring Machine (**CMM**) is a reliable and robust tool for obtaining measurements of micrometer accuracy on machine products [5]. The basic types of CMM are moving bridge, gantry and horizontal arm which reflect the arrangement of measuring arms in Cartesian (x,y,z) Coordinate system. The elementary components of a CMM are a table for workpiece mounting, the mechanical arms, the probe for component location, contact/measuring scales to determine the component position and the power drive systems. The CMM performance is best assessed in terms of the speed and accuracy of measuring a point on the workpiece [7, 4]. The demand for product quality and fast inspection tools to reduce manufacturing cycle time has continuously grown in recent years. The conventional CMMs can no longer match the modern and computerised production lines. In parallel the advent of high speed computing machines and the development of high power electronic components, which can replace the mechanical and analogue elements, have now made it possible to design high speed and very accurate CMMs. The controller design to fully utilise the power of these new machines are a major challenge since the two conflicting requirements of speed and accuracy should be simultaneously satisfied [6].

The (CMM) considered in this paper consists of three orthogonal arms in the x,y and z axes as shown in figure 1 and a two DOF probe capable of rotating 180° about the horizontal axis and 0° to 105° about the vertical (figure 1-b). The arms are controlled by independent servo loops consisting of a PWM current amplifier, three-phase brushless DC motors with incremental shaft encoders, a preamp component for signal conditioning and velocity loop control and a position controller implemented in software and downloaded to a transputer. The position transducer is a combination of an optical reading head and grating. The output of the transducer is fed to a transputer for decoding the position.

1.1 CMM Operation

The top level control is provided through a computer. A target position, which is a small distance from the actual point to be measured on the workpiece, is supplied to the computer. The probe is then moved to this target position at the highest acceleration and velocity possible. From this target position the probe is then moved at a slow constant speed until contact is made with the workpiece. A triggering mechanism in the probe sends an interrupt signal to the system which rapidly halts all the movement of the servos.

The movement of the probe from the current position to

the target position can ideally be performed in minimum time by driving the motor with its highest acceleration until maximum speed is reached. The motor is run with this speed until the probe is close to the target position. The motor is then stopped by applying maximum acceleration in the reverse direction. To avoid wear on the system, step inputs are unacceptable. Also maximum acceleration (6000 mm/sec^2) and velocity (400 mm/sec) are limited by motor and amplifier performance specifications. To overcome these restrictions, a parabolic acceleration profile is being used in practice to calculate the reference position trajectory. This is then sampled and the intermediate values at each sampling instant are applied to the command input of the position servo loop.

The success of the above high level control strategy depends on the performance of the low level control schemes, i.e the velocity and position loops in each axes. To ensure zero position error, the velocity loop should be about 10 times faster than the position loop. No velocity overshoot is allowed since this may drive the amplifier to saturation and possible limit cycle oscillation may occur. For safety reason, the position loop response is also required to be critically damped. Moreover, the position controller should be robust to any changes in the velocity loop parameters and load changes.

This paper is concerned with developing a design environment for CMM control design. To this end, a demonstrator test rig is constructed which consists of the basic elements of the servo loop, i.e actuator, sensor, CMM arm and the controller. A nonlinear model of the system is developed and implemented on an interactive nonlinear simulation package (EASY5) [13]. This model is linearised to develop simple models for control design studies. The model is validated using raw data generated from the test rig. An iterative design approach is adopted to cheaply and efficiently optimise the system performance in terms of speed and positioning accuracy of the probe.

2 Modelling

The nonlinear models of the servo systems, the arms and the probe are independently developed and then combined to build the overall model of the system. These models are modular and may be easily modified to incorporate any change in the system design. Simulation results are validated against the experimental data obtained from the test rig.

2.1 Servo Loop Modelling

Figure 2 shows the velocity loop, containing preamp, amplifier, motor and velocity transducer. The derivation of the preamp model, which contains the current and velocity loop control electronics, is discussed in [7]. Only, the reduced model used for control design is presented in figure 2. Three potentiometers of gains k_1, k_2, k_3 are provided on the preamp board to tune the velocity loop. The preamp is designed to have a two degree of freedom control structure in which the set point is controlled through G_c and G_L. The velocity feedback signal is controlled by G_f and G_L. This control structure helps to maintain good tracking as well as loop stability. Note that the set point is not the conventional step changes but a profile similar to a Sigmoid function designed to maximise the speed of the probe positioning system. The velocity controller is tuned using the three gains k_1, k_2, k_3 to adjust the frequency responses of the three networks G_c, (lead/lag), G_f (lag) and G_L (lag), respectively. The amplifier employed in each servo loop is a PWM device, the output being a constant current pulse of duration proportional to the magnitude of the input voltage [2]. The PWM amplifier operates with the power transistors either off or (almost) saturated thus giving a lower power consumption when compared with an equivalent linear amplifier. The sampling frequency of the amplifier is 16 kHz. Since the system bandwidth is less than 100Hz, output ripple is negligible. The amplifier is modelled as a first order lag since the system acts as a low pass filter [2] [7]. Three three-phase brushless servomotors are used to drive each arms. Since the amplifier output is current [16], the motor's electrical time constant is negligible compared to the other system time constants. The motor is therefore modelled as a first order lag [7].

2.2 The CMM model

Figure 1-a shows a vertical column CMM. It consists of three perpendicular arms for the movement in the Cartesian coordinate system.The hollow box construction of the arms helps to maximise the stiffness to mass ratio which is an important factor affecting the measuring accuracy [13]. The probing system consists of the probe head, probe and stylus (figure 1-b). The probe is controlled by two stepper motors. For good measurement repeatability, the probe has only two DOF. Thus the arms and the probe together have five DOF. In addition a rotary table provides an extra DOF for more flexibility A mass counterbalance is incorporated in the vertical, z, arm to compensate the gravity induced force.

For modelling purposes, the arms and the probe are considered as a set of rigid links connected together in series by three prismatic and two revolute joints. There are as many joints as the number of DOF, i.e 5. Using a systematic method proposed by Denavit and Harterberg [10], base and local Cartesian coordinate frames for each link are defined as shown in figure 1-a. The variables q_1, q_2, q_3 represent the linear and q_4, q_5 angular displacements for each joint. Knowing the transformation matrices for two adjacent coordinate frame, the kinematic equations between the position and orientation of any point on the CMM and joint variables can be established. Similarly, the kinematic equations between the velocity of joint variables and rotational velocities of the links may be formulated. Finally the acceleration of links may be derived by differentiating the expressions for translational and rotational velocities. The Euler-Lagrange's equations [10] commonly used to obtain dynamic models for robots can be applied to CMM. The five equations derived in [8] is presented here.

$$
\begin{aligned}
T_{L1} &= g_1 m_5 l_{c5}(-S_4 C_5 \ddot{q}_4 - C_4 S_5 \ddot{q}_5 + \\
& \quad 2 S_4 S_5 \dot{q}_4 \dot{q}_5 - C_4 C_5 \dot{q}_4^2 - C_4 C_5 \dot{q}_5^2) \\
T_{L2} &= g_2 m_5 l_{c5}(C_4 C_5 \ddot{q}_4 - S_4 S_5 \ddot{q}_5 - 2 C_4 \\
& \quad S_5 \dot{q}_4 \dot{q}_5 - S_4 C_5 \dot{q}_4^2 - S_4 C_5 \dot{q}_5^2) \\
T_{L3} &= g_3 m_5 l_{c5}(C_5 \ddot{q}_5 - S_5 \dot{q}_5^2) \\
T_{L4} &= m_5 l_{c5}(-S_4 C_5 \ddot{q}_1 + C_4 C_5 \ddot{q}_2) \\
& \quad + (I_{224} + (m_5 l_{c5}^2 + I_{225}) C_5^2 + \\
& \quad I_{115} S_5^2 + 2 I_{125} C_5 S_5) \ddot{q}_4 + (I_{135} \\
& \quad S_5 + I_{235} C_5) \ddot{q}_5 + 2(I_{115} - I_{225} - m_5 \\
& \quad l_{c5}^2) C_5 S_5 + I_{125}(C_5^2 - S_5^2)) \\
& \quad \dot{q}_4 \dot{q}_5 + (I_{135} C_5 - I_{235} S_5) \dot{q}_5^2 \\
T_{L5} &= m_5 l_{c5}(-C_4 S_5 \ddot{q}_1 - S_4 S_5 \ddot{q}_2 + C_5 \\
& \quad \ddot{q}_3) + (I_{135} S_5 + I_{235} C_5) \ddot{q}_4 + (m_5 l_{c5}^2 \\
& \quad + I_{335}) \ddot{q}_5 - (I_{115} - I_{225} - m_5 l_{c5}^2) C_5 S_5 \\
& \quad + I_{125}(C_5^2 - S_5^2) \dot{q}_4^2
\end{aligned}
\quad (1)
$$

where, T_{Lk} : load torque of link k, m_5 : mass of link 5, q_k : joint variable k, l_{c5} : distance of center of mass of link 5 from joint 5, g_k : gear ratio of link k, $C_k = \sin q_k$, $S_k = \cos q_k$, $I_{ijk} = (i,j)th$ element of inertia matrix of link k.

2.3 Probe Model

The probe is situated at the end of arm 3. It mainly consists of a motorized probe head and a touch-trigger stylus. It is also possible to use a probe extension to lengthen the probe. It is assumed that both stepper motors have permanent-magnetic rotor with 2 phase excitation. Each stepper motor can then be described by the following nonlinear differential equations [9]:

$$
\begin{aligned}
\dot{I}_a &= (V_a - R I_a + K_m \omega \sin(N_r \theta))/L \\
\dot{I}_b &= (V_b - R I_b + K_m \omega \cos(N_r \theta))/L \\
\dot{\omega} &= [K_m(I_b \cos(N_r \theta) - I_a \sin(N_r \\
& \quad \theta)) - B\omega) - T_L]/J_s \\
\dot{\theta} &= \omega
\end{aligned}
\quad (2)
$$

Where L: self inductance of each phase winding, R: resistance of each phase winding, K: motor torque constant, N: number of rotor teeth, J: rotor inertia, B: viscous friction constant, I_a, I_b are the currents in phase A and B and ω and θ are are the angular velocity and position. T_L is the load torque given by equations 1. The inputs to the stepper motor are the discrete phase voltages V_a and V_b which determine the steady state position and the phase currents.

The CMM and probe models given by equations 1 and 2 are combined and linearised around a nominal operating point. This model in conjunction with the stepper motor model give the total linear model of the system.

3 Simulation Studies

Simulations have been carried out to investigate the probe-arms interaction dynamics. The velocity loops are tuned to obtain good transient response behaviour. This study demonstrates that the interactions on the probe come only from the accelerations of the arms as expected and such interactions are significant (maximum deviation of about 1.5 degrees from the nominal probe positions). However, the interactions from the probe to the arms are very small. Thus the effect of the load torques on the arm dynamics may neglected.

The CMM/probe model has been linearised about nominal probe positions. Comparisons of the responses of the non-linear and linearised model show great accuracy in the linearised model and this suggests that application of linear control theory is possible. Furthermore, it can be seen from the linearised model that the dynamics of the two stepper motors are decoupled, hence they can be treated independently. Finally, the order of the stepper motor model can be further reduced with the loss of virtually no accuracy.

4 Sensor Design and Test Rig

A one axis test rig has been constructed in order to test and develop the control algorithms. This includes an Electrocraft servo system [1] consisting of preamp, amplifier and servomotor and a low cost position transducer expected to give accuracy of 0.05 mm. The motor is connected to a worm screw and a trolley is mounted upon this to give linear position. The arm is modelled by a flexible beam mounted on the worm screw.

The position transducer is based on Moire Fringe techniques. An optical grating with a resolution of 20 lines/mm is traversed by a reading head with two light sensors, at a constant height of $0.2mm$. This height is maintained by using compressed air to support the CMM axes and also to provide a frictionless servo. The two out of phase sine wave outputs from the reading head may be fed to a transputer bank to resolve the position to 0.1 μm.

It is not possible to implement this transducer on the servo test bench at present, due to difficulties in maintaining the head level above the optical grating. A transducer with lower resolution has therefore been developed which simply counts every line crossed by the reading head. Experiment has shown that the outputs of the reading head are bipolar but without guaranteed magnitude. These outputs are amplified by an inverting OP amp circuit to levels compatible with Transistor Transistor Logic (TTL). The bipolar signals are then rectified and fed to a Schmitt inverter which provides a square wave output of the reading head signals. The two out of phase signals are then decoded for direction by feeding one signal to the D input of a D-type flip flop and the other signal to the clock input. Further decode logic dependent on the type of counter used is required to count the pulses in the correct direction.

The track length is 1.2m and this requires 24000 counts which is covered by 2^{15}. Thus a fifteen bit counter is required. Two eight bit input ports of a Direct Digital Interface (DDI) are used to read this value. Direct connection of these lines to the computer (IBM-PC) has proved difficult due to interference and cross-talk corrupting the count. A temporary solution has been to use further decode logic and count every fourth line giving a 12 bit counter with a resolution of 0.2mm and a track length of 0.819m. The circuit diagram for this device is shown in figure 3.

The basic interface between the servo and IBM-PC requires a Digital to Analogue Converter (DAC) to provide the velocity reference signal, an Analogue to Digital Converter (ADC) to return the velocity feedback signal to the computer and a Direct Digital Interface (DDI) to control and read the position transducer. These requirements are met by the PC-30A, a commercially available Input/Output (I/O) card. This card offers a 16 Channel ADC, two 8 bit DAC and 12 bit DAC which are controlled through a Programmable Peripheral Interface (PPI) chip (8255). A second PPI is used to interface with the position transducer programmed as two 8 bit input ports for reading position and an 8 bit output port used to initialise the position transducer. The software to control the interface is written in 'C'.

5 Control Design

The main control design objective is to minimise the inspection time while meeting common design requirements such as zero overshoot, zero steady state error and stability robustness. The linear model employed for control design is a three inputs four outputs transfer function matrix which may be transformed to a 14th order state space model. The controlled variables are the Cartesian probe positions. The measured variables are the servo position and velocities in each axes. The reference input is a position profile which is designed to achieve maximum acceleration and deccelatrion in minimum time. Due to the complexity of the reference input, it is essential to use a two degrees of freedom controller to meet both the performance and stability requirements. Unlike the conventional control problems where the controlled variables are also the measured variables, the probe positions which are ideally to be controlled are not measured [15].

As the effect of the probe on the arms dynamics is negligible, an individual channel control design may be adopted to reduce the multivariable problem to three single input single output control design problems. The interaction effects are considered as disturbances and attempts have been made to minimise the effect of loop interactions as much as possible. In order to keep the controller as simple as possible, its structure is restricted to simple PID and Lead-Lag compensators. This is essential for the high speed machine considered here since the control input calculation should be performed in a fraction of microseconds.

The interaction between the control design and the process is performed through nine potentiometers which regulates the current and velocity loops. These parameters are initially set to give smooth velocity responses when the position loop is open. When the position controller is designed and implemented, these potentiometers are retuned for optimal performance.

5.1 Velocity Loop Optimisation

The velocity control loop should operate about 10 times faster than the position loop. This is to ensure that the position controller output which is supplied as a set point to the velocity loop is reached to the correct level within the sample time ($1 msec$). This implies that the velocity controller should be designed such that the settling time is minimised.

As the motor mostly operates with maximum speed, only small overshoot is allowed. This requires that the velocity loop bandwidth should be maximised for optimal operation and also for rejecting any disturbances which may cause error in probe positioning. The velocity loop is designed such that the potentiometer gains can only be changed manually. Therefore, a trail and error design procedure based on the frequency response is followed to optimise the velocity loop.

The open loop system response including the preamp controller is shown in figure 4. The effect of varying the preamp gains is clearly visible from the graphs and is summarised in table 1.

k_1	k_3	GM	PM	BW
.01	1.0	7	14	1000
0.5	1.0	31	17	200
1.0	1.0	37	13	100
1.0	0.083	44	49	100
1.0	0.5	40	20	100
0.05	0.5	17	37	100

Table 1 variations of Phase Margin(PM), Gain Margin (GM) and BandWidth(BW)

Table1 and the frequency responses of figure 4-a demonstrate that increasing k_1 increases the the gain margin, thus increasing system stability, while the phase margin is not significantly altered. Variations in k_2 affect both the gain and the phase margins. An increase in k_2 results in a decrease in gain and phase margins. Figure 4-a also indicates that k_2 introduces a phase lead in the high frequency region and the amount of phase lead increases as k_2 decreases. This introduces derivative action at high frequencies which is not desirable for noise rejection. The high frequency roll-off rate is independent of the tuning parameters and has a value of 30 dB/decade. The closed loop time domain responses are shown in figure 4-b, illustrating the effect of changing these parameters. These indicate that increasing k_1 increases the settling time and make the response oscillatory as indicated in Table2. k_2 is seen to affect the closed loop gain of the system but not the settling time. As k_2 is decreased the system gain increases. An increase in gain of the order of 10 is apparent as k_2 is varied from 1 to 0.083. A tuned response can be obtained giving a critically damped response and a bandwidth of approximately 100 Hz.

k_1	τ_s	M_p
.01	0.04	11
0.5	0.15	60
1.0	0.4	74

Table 2 Settling time τ_s and overshoot M_p

5.2 Position Loop Optimisation

The design of the PID controller for the position loop of the servo system is based on moving a point on the uncompensated Nyquist curve to a desirable position so that the control design specifications are satisfied. The main objective is to minimise the probe positioning time between two points in the Cartesian space.

The position response should be critically damped to avoid saturation and possible damage to the probe in confined spaces. This implies that the loop should have a phase margin of about 60°. It is also necessary to maximise the bandwidth of the loop such that the effect of disturbances such as arm vibration [12] can be compensated. This will however decrease the gain at lower frequency and introduces steady state error. A trade-off between these specifications is obtained by iteration.

Having specified the phase margin θ_s and bandwidth ω_c, two points on the Nyquist curve are determined as follows:

$$G_p(j\omega_c) = r_p \exp j(\pi + \theta_p)$$
$$HG_p(j\omega_c) = \exp j(\pi + \theta_s) \quad (3)$$

Where G_p is the transfer function of position to the velocity set point, and $G_c = K(1 + 1/T_i s + T_d s)$ is the position controller. The PID gains may be calculated using the relationships given in [11], i.e:

$$K = (\cos(\theta_s - \theta_p))/r_p$$
$$T_d = \tan(\theta_s - \theta_p)/\omega_c + 1/(\omega_c^2 T_i)$$
$$T_i = \sigma T_d \quad (4)$$

Where σ is a tuning parameter to determine a trade-off between the effect of integral action and derivative action.

The uncompensated and compensated Nyquist curves for different bandwidth are shown in figure 5. This shows that as the bandwidth increases the low frequency gain decreases and high frequency amplification increases. The position loop time responses are shown in figure 6. These two figures can be used to find the most appropriate controller parameters for the position loop.

6 Experimental Results

The position controller designed in previous section is implemented on the test rig to investigate the performance of the servo loop. The step responses of the position, velocity and the acceleration are shown in figures 7 and 8. The position response has a small overshoot which indicates the controller has high proportional gain. Fine tuning of the controller is therefore required to reduce the overshoot. Reducing the proportional gain will reduce the overshoot but the rise time will also decrease as shown in the set of responses in figure 7. These experiments also illustrate that the closed loop response is sensitive to the magnitude of the position loop set point. Thus, an adaptive [14] controller is needed to ensure that the design specification are met for a wide range of operating conditions.

7 Conclusion

An integrated process and control design environment is proposed for design and control loop optimisation for fast Coordinate Measuring Machine. The nonlinear and linear models are developed and implemented on EASY5 [13] which provides a cheap and accurate tool for simulating complex dynamical systems when the structure and the parameters of the model are known. The linear model is used to optimise the velocity and position loops. The performance of the controller is evaluated on the nonlinear model before implementation on the real system. By several design iterations, a controller can be found for the three axis of CMM. The controller is designed such that the speed and accuracy is maximised. Experimental results are presented to show the effectiveness of the proposed scheme.

8 Acknowledgement

The authors are grateful to ACME directorate and Ferranti Industrial Electronics Ltd for their support for this project.

References

[1] Electro-craft Limited,' D.C. Motors, Speed Controls', Servo Systems.*Technical Manual, 1990*

[2] R J Sandor and R Unnikroshnan,' PWM Motor Control: Model And Servo Analysis.',*IEEE conf record of Industry applications society annual meeting, Chicago,1984.*

[3] R D Klafter, T A Cmielewski and M Negin,' Robotic Engineering: An Integrated Approach.,*Prentice-Hall, 1989.*

[4] A Harvie,'Factors affecting component measurement on Coordinate Measuring Machine',*Precision Engineering, Vol8, No 1, 1986*

[5] K Lenz and W Merzenich,'Achievement of accuracy by error compensation of large CMMs',*Precision Engineering, Vol10, No 4, 1988*

[6] A T Sutherland and D W Wright,'Optimising a servo system for coordinate measuring machine.',*Precision Engineering, Vol 9 , No 4, 1987.*

[7] J McIntyre, M R Katebi and M J Grimble,'Modelling and simulation of a Co-ordinate Measuring Machine,*ICU report 339,September 1991.*

[8] T Lee, M R Katebi and M J Grimble,'Nonlinear and Linear models for Co-ordinate Measuring Machine,*ICU report 340,September 1991.*

[9] T Kenjo, ' Stepping motors and their microprocessor controls ,*Oxford,UK,Clarendon,1984.*

[10] A J Koivo,'Fundamentals for control of robotic manipulators ',*John Wiley, 1989*

[11] K J Astrom and T Haggland,'Automatic PID regulators'.

[12] M R Katebi,'Control design for flexible structures', *1st European Conf on Smart Structures, Glasgow, May 1992*

[13] ,'EASY5 Engineering Analysis System',*The Boeing Company, , 1987.*

[14] H Chuang and C Liu,' A model-referenced adaptive strategy for improving contour accuracy of multiaxis machine tools', *IEEE Trans. on Industry Applications, vol 28 No 1, 1992.*

[15] M J Grimble, 'Polynomial solution of standard H_2 optimal problem for machine control system applications',*Industrial Control Centre Report, 337, 1992.*

[16] Y Dote, ' Servo motor and motion control using digital signal processors',*Prentice Hall and Texas Instruments, Newyork, 1990.*

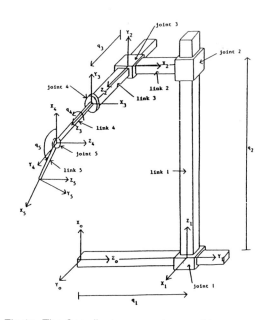

Fig 1a The Coordinate measuring machine

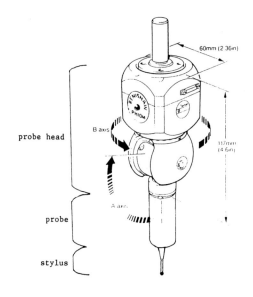

Fig 1b The probe

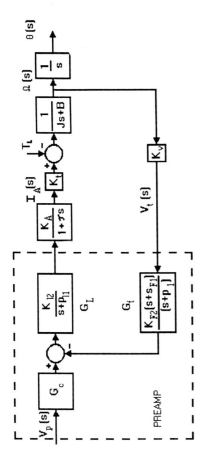

Fig 2 Velocity loop block diagram

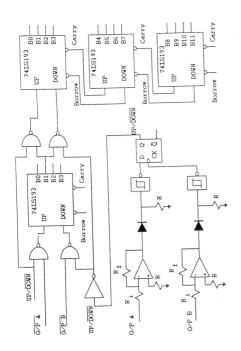

Fig 3 The transducer circuit diagram

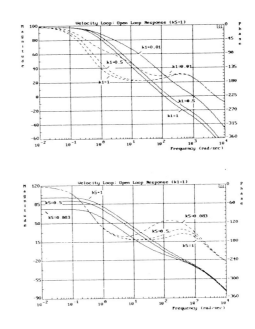

Fig 4a The Frequency responses

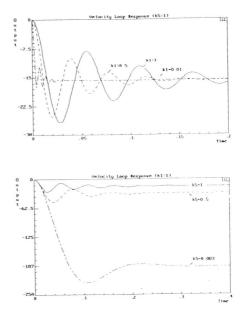

Fig 4b The time responses

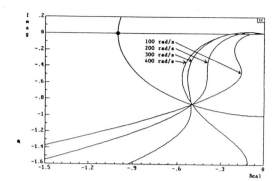

Fig 5   The Nyquist plots for different bandwidth

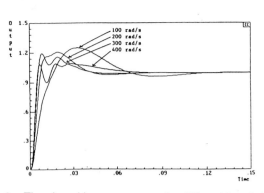

Fig 6   The closed loop responses for different bandwidth

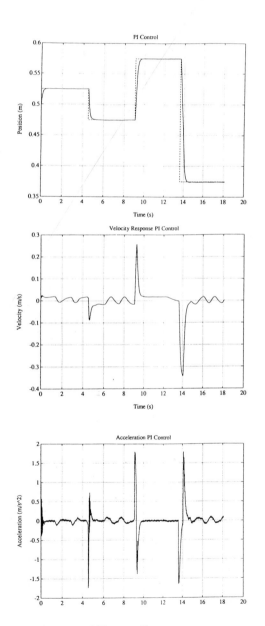

Fig 7   The fine-tuned PI controller

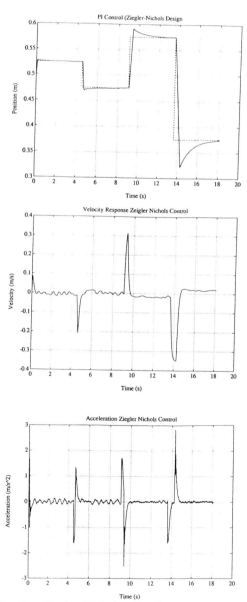

Fig 8   The Ziegler–Nichols controller

Camshaft profile measurement – a mechatronic approach

M R JACKSON, BSc, MSc, PhD
Loughborough University of Technology, UK

SYNOPSIS A mechatronic approach to the development of a low cost cam profile digitiser is described. The work details the elements of the mechanical arrangement, the transducer requirements and selection, interface scheme for an IBM-Personal Computer and the software functions developed in the 'C' programming language. The performance of the instrument is reviewed and further development discussed.

1 INTRODUCTION

Cam profile measurement is required primarily to ensure conformance between design and manufactured profiles, and secondly for cam profile identification during machinery performance evaluation. Measurement of cam forms is not a new requirement and in fact there are a number of expensive machines which may be used for assessment of profiles (1). In some applications the error between the desired and actual profile is fed back to the cam grinding machine, sometimes as adaptive control, in order to improve profile quality (2).

The instrument described in this paper is one approach to cam profile measurement and has been undertaken as a background task sometimes involving student project work. The primary aim is to develop a low cost instrument which can digitise cam profiles and display various parameters for cam form verification. The ideas developed will provide a basis for an extension to the design to provide a higher resolution on the radial transducer to enable assessment of cam form variation. Further work includes the development of portable equipment for in-situ verification of cam profiles and the development of a fully automatic cam profile measuring machine. A secondary aim of the work is to provide a 'lid off' product which can be used for teaching in mechatronics and related studies. The work does not attempt to develop an instrument with the very high levels of accuracy of the expensive equipment nor does it aim to measure surface texture of the cam profile.

This work is limited to the investigation of automotive and plate type cams where angle and radius are the digitised and manipulated parameters. The principles described here could be applied with some modification to the form measurement of barrel cams.

2 INITIAL SPECIFICATION

Maximum cam radius: 30 mm
Stroke of radial transducer: 20 mm
Angular Resolution: 0.1 degrees
Radial Resolution: 0.01 mm & 0.001 mm
Low Cost
Graphical User Interface
Database of Cam Profiles
Cartesian Output for parameters of interest
Geometric Output to allow visual comparison of two or more profiles
Easy to use

3 MECHANICAL ASPECTS

High quality inspection machines as detailed in (1) use air bearings or similar high quality supports to rotate the component under test. In this way an independent assessment of component quality can be made. This approach accounts for a major factor in the machine cost as the bearing and fixturing elements are made to very high quality in order to achieve the levels of accuracy and precision in the measurement process.

Most cam profiles are produced on machine tools where conical centres are used to locate the camshaft on a reference axis. These centre holes in the ends of the camshaft are utilised on the cam profile measuring instrument in order to locate the camshaft on the reference axis and allow the cam profile to rotate past the stationary radial transducer; the method is well known in the inspection of shafts and is adequate for the accuracy outlined in the specification.

Since automotive cams are slender items the axis of rotation of the camshaft is arranged in the vertical plane to minimise the effects of camshaft flexure due to gravity and asymetry.

The preferred method of locating a camshaft between centres is to use two fixed centres. Whilst this approach is best for precision it presents some difficulties with the angular transducer location. The problem is caused by the necessity to transfer the angular rotation of the camshaft to the rotary transducer whilst maintaining a fixed (i.e. none-rotating) centre. The two primary solutions to this are: (a) to locate a rotary transducer adjacent to one of the centres and link it to the camshaft by an endless belt and pulley arrangement or by a friction wheel running on one of the cam bearing surfaces, or (b) design the centre so that the drive from the camshaft to the transducer and the transducer hardware are coaxial with the fixed centre. Method (b) is a more elegant solution for precision of measurement and operational reasons, the disadvantage is potentially one of cost and difficulties with in service replacement of the transducer elements. Neither of these two problems are insurmountable and will form part of a future development. As a compromise for the first prototype a rotating centre is used at the base of the machine as shown in Figure 1. This arrangement makes use of the shaft and bearings of a heavy duty digital encoder design which has a static axial loading capacity of 10 kg.

The eccentricity at the rotating centre (typically 0.015 mm) is measured during assembly and then compensated for in software at the different cam lobe measurement positions along the vertical axis. This approach is possible because the variation in eccentricity due to bearing effects in the encoder is not significant for the level of precision in the specification and secondly, because the user enters the linear distance of each lobe from the bottom face of each type of camshaft.

In keeping with the low cost nature of this device the cam under inspection is rotated by hand. The system is designed to tolerate variations in rotational speed during the digitising cycle and indicates if the speed exceeds the upper limit of one revolution per second maximum angular velocity. If this is the case the cycle must be repeated at a lower speed.

The means of rotating the camshaft and driving the transducer to provide feed back of angle is achieved as shown in Figure 2. For the case illustrated the existing drive pin of the camshaft is utilised as the reaction point with the drive tube fixture and the rotating centre mounted on the encoder shaft. The end of the drive tube adjacent to the drive pin is slotted to allow the pin to sit within the tube wall. The operator then adjusts the ball-end set screw to react the pin against the opposite ball-end screw in order to remove any clearance in the direction of rotation and provide a flexible joint for the remaining degrees of freedom. This approach works very well for the camshaft type illustrated, however, purpose made drive tubes are required for variants of drive pin diameter and position. Alternatively, a universal fixture arrangement can be used but this tends to result in hardware which is somewhat cumbersome in this situation. Further work is required to develop a solution which accomodates a large number of cam types with minimum special machining.

4 SENSOR ASPECTS

The cam profile measuring instrument is to be interfaced to a microcomputer for the data logging, storage and display functions. Generally, digital sensors are easier to interface to computing equipment than analogue sensors, and the software requirements are less demanding. Bearing this in mind the choice of angle of rotation sensor is mainly about absolute or incremental digital encoders. An angular incremental encoder is used as an angle of rotation sensor because of relatively low cost, high resolution and signal stability. The encoder selected is a Hohner 3000 series heavy duty type with a line count of 3600 per revolution, with two channels, allowing 14400 quadrature counts per revolution, cost approximately £300.

This heavy duty type of encoder is necessary as the weight of the camshaft and any mechanical shocks due to the loading of the camshaft must not cause deterioration of the bearings and the encoding function.

The software routines described later allow a scaling factor to be introduced to evaluate the effect of digitising resolution on data presentation and hence profile assessment. The software sampling routine limits the maximum speed of rotation of the encoder to one revolution per second determined by the interface and software requirements. With a one revolution per second maximum angular velocity the encoder quadrature count frequency is 14.4 kHz which is within the specification for this encoder.

The specification for the radial transducer is listed as 20 mm stroke with a resolution of 0.01 mm and 0.001 mm. The reason for two resolution values is because cam form verification can generally be achieved with 0.01 mm resolution, whilst variations in form quality require a higher resolution device (i.e. 0.001 mm). There are a number of linear digital encoders available with resolution of the order of one micrometre. Whilst these devices meet the resolution required for quality measurement, the cost of these types of device is correspondingly high (typically £500 -£1000) for 25 mm - 50 mm stroke). A lower resolution, lower cost radial transducer option is desirable for use in cam form verification and whilst there are low cost linear encoder kits available with 0.07 mm resolution these are clearly inadequate. An alternative approach using a Linear Variable Differential Transformer (LVDT) transducer is used in the prototype form verification instrument. A 25 mm stroke unit (ACT500A) complete with amplifier (S7AC) costs approximately £200 (from RDP Electronics) with a linearity of 0.1% full scale and a signal to noise ratio of 4000. The range of the radial measurement is 20 millimetres with a resolution of 0.01 millimetres, producing a range to resolution value of 2000. The LVDT arrangement is therefore adequate for this system on a range to resolution assessment.

However, the radial transducer must be supported to accept the side loading of the cam on the follower. Linear transducers do not generally accommodate side loading and so a modified unit has been developed for experimental purposes (Figure 3). The LVDT shaft is extended through a linear bearing which supports the side loading whilst allowing the cam follower to move freely in the radial direction under the action of the light spring force (typically 50 grammes) from the LVDT unit. A precision roller is used as the cam follower to reduce lateral forces on and deflection of the LVDT extension shaft. The radius of the roller is 5 millimetres which is compensated for using software to produce the true cam profile (3).

5 USER INTERFACE

The desired output from the measuring instrument is in the form of profiles and graphs which suggests the use of a Personal Computer (PC) since hardware graphics options for this type of machine are of reasonable cost and programming languages such as Turbo C provide graphics functions. Lap top computers or other lower cost devices can be used with corresponding reduction in quality and speed of graphics presentation.

The user interface for the first prototype is menu driven prompting the operator to digitise cam lobe numbers in ascending order. The operator can write the matrix of cam lobe profile data to disc file and similarly retrieve data from file. The data may be displayed as an enlarged cam profile on the PC Visual Display Unit (VDU), as a table of angle and radius on the PC printer or as graphical output of angle and corresponding radius on the PC VDU.

6 TRANSDUCER - PC INTERFACE SCHEME

A schematic of the interface scheme is shown in Figure 4 where a specially designed printed circuit board (PCB) is used to provide data handling between the two transducers and the PC system bus.

The interface scheme for the encoder and the LVDT utilises two principal integrated circuits. These are the Hewlet Packard HCTL-2000 decoder/counter and the Analogue Devices AD574 Analogue to Digital Converter (ADC). The HCTL-2000 is a 12 bit counter with quadrature decoding and digital filtering built into the package. Data transfer from the 12 bit counter register to the PC system bus is in the form of two single bytes with leading zeros for the most significant (MS) nibble. This eight bit data transfer approach ensures that a wide range of PC type machines can be used. The HCTL-2000 12 bit counter rolls over within a third of a revolution of the encoder due to the high encoder resolution, therefore the data logging routine is designed to take account of this fact by forming an absolute reading in memory for the angle of rotation. The speed of execution of this routine is the limiting factor on operating speed of rotation of the encoder. Whilst this present limit does not affect the normal operation of the instrument, a reduction in encoder resolution would result in an increased upper limit of rotational speed.

The AD574 is a 12 bit ADC of the successive approximation type with a conversion time of typically 25 microseconds. The choice of a 12 bit converter is arrived at by consideration of the amplifier signal range (+/- 10 volts) and the quoted electrical noise level in the amplifier output lines (typically 5 millivolts peak to peak). An 8 bit ADC will provide a quantisation value of 78 millivolts which translates to 78 micrometres resolution on the radial transducer. This is clearly inadequate for the target specification. A 10 bit ADC provides a quantisation value of 20 micrometres, a 12 bit ADC provides 5 micrometres which is of the same order as the maximum quoted noise level for the amplifier. A 12 bit converter is chosen on the basis that the 10 bit is too coarse and clearly a 16 bit converter will be down in the noise ripple.

Conversion is started by a write instruction to a specific address and a polling routine is used to test the device status pin via the PC data bus and hence establish end of conversion. The 12 bits of ADC data are transferred to the PC bus as two one byte reads with the least significant (LS) nibble of the lower byte consisting of trailing zeros, this latter point is taken into account when writing the system software.

Additional integrated circuits include two gate array logic devices (GALs) which are used with a software design package called CUPL to implement the address decoding for the two principal integrated circuits and some additional functions such as clock divide and ADC status pin buffering to the PC bus.

The prototype interface card is of wire wrapped construction, estimated costs of a production PCB are £100-£150 including all integrated circuits, assembly and test.

7 SOFTWARE DESIGN & IMPLEMENTATION

The 'C' programming language is used for this application because of it's suitability for general mechatronics ('C' is a 'middle level' language).

Figure 5 shows the structure of the 'C' functions written for this application. The main function is menu() which provides three options which in turn call alternative functions. Cam_digitise() checks for the encoder reading incrementing by the scale count (typically 10 counts) and then uses two transducer specific functions get_angle() and get_radius() together with write_to_disc(). Get_angle() which reads the HCTL 2000 in two bytes, forms the absolute angle of rotation in 32 bits and returns this angle. This approach means that a consistent speed of rotation is not required since the array index pointer is updated only when a set angular increment is achieved.

Get_radius() reads the AD574 in two bytes, applies a scale factor and returns the actual radius. After 360 degrees of rotation from the start point of data logging cam_digitise() will prompt for the next lobe and the functions repeat until the operator quits this part of the cycle. The data for angle and radius is stored in two arrays for each cam lobe number. The function write_to_disc() allows a user designated filename for storage of the complete cam data on hard or floppy disc for later retrieval.

The function graphical_output() takes the user filename (or current data) and uses display_form() to produce a display of the enlarged cam form on the PC VDU. The function plot_curve() is used to produce the display of radius vs: angle. A table of results is sent to the printer using the functions shown.

8 MACHINE OPERATION

The operator locates the camshaft and drive peg in the rotating centre assembly and then positions the fixed centre to secure the camshaft between the two centres. The clearance between the camshaft pin and the drive tube is taken up by adjusting the ball end set screws. The LVDT transducer slide is moved onto the first cam lobe of interest and the camshaft is rotated by hand at reasonable speed until one revolution is exceeded (Figure 6). If the speed of rotation of the camshaft is too high the software will indicate an error and the lobe digitising cycle must be repeated.

The remaining cam lobes on the camshaft are digitised starting each lobe cycle at any angular position. This feature is produced by always maintaining a count of the absolute angular position over several revolutions of the camshaft using the 'C' routine cam_digitise(). The data may be written to disc file and then retrieved for comparison at a later date.

9 RESULTS

The results obtained during initial testing of the machine have been very encouraging. The LVDT and amplifier were tested seperately using slip gauges in order to establish the actual transducer constant. This was found to be 0.8 millivolts output per micrometre input displacement of the LVDT, and found to be linear within 0.1% with no appreciable drift over 24 hours. The input signal conditioning, the ADC and software were tested using a precision voltage source to confirm linearity and establish the actual quantisation value of 4.882 millivolts per decimal unit. This determines the practical resolution of the LVDT and interface as 6.1 micrometres which is better than the 0.010 mm target value for cam form verification.

The encoder interface was checked using external counters to establish correspondance between the number of counts generated and the values read by the function get_angle().

The speed of digitising is typically five seconds per cam lobe which is more than adequate for the operation of the prototype. It is worth noting that since this instrument was developed Hewlet Packard have updated the HCTL-20xx series to include a 16 bit counter which can be cascaded with conventional counter integrated circuits in order to construct 24 bit counters (for example). This would relax the software timing requirements for the encoder read routine and allow higher speed of operation in other applications.

Figure 7 shows an example of the graphical output obtained for a single cam lobe. A full 360 degrees of rotation is shown with the radius displayed relative to the base circle of the cam profile in order to increase display magnification. This particular cam is from a Moto Guzzi SP 1000 motorcycle. Tabular output and cam profile display are also provided.

Additional software is required to enable manipulation of adajacent cam lobe profiles in order to establish information such as valve timing overlap, velocity and acceleration data and error from the desired form.

More test work on reference cams and comparison with more expensive equipment is planned.

10 CONCLUSIONS

This work demonstrates that it is possible to achieve a low cost easy to use cam profile data logging and display facility providing adequate performance. Additional software development will provide a cam profile analysis facility.

The teaching experience provided by this device is:

Mechatronic philosophy

Transducer selection,

PC bus interfacing of 12 bit ADC

PC bus interfacing of 12 bit decoder counter,

'C' programming

Manufacturing Engineering.

11 FUTURE WORK

Further work will provide additional data such as error between cam profiles, first and second derivatives of radius with respect to angle and a graphical display of cam profile showing exaggerated differences between profiles.

A further refinement is possible to eliminate the use of a rotating centre by constructing a fixed centre incorporating a modular encoder.

A fully automatic low cost system is possible using additional software for the PC written in'C' to control three d.c. motors and encoders via individual low cost servo controllers such as the Hewlet Packard HCTL-1000. An automatic measuring cycle for each different type of cam will allow high speed assessment of cam profiles. The servo axis for the radial motion will serve two functions 1) to retract the follower to a safe position during vertical axis indexing to cam lobe digitising positions and controlled positioning of the follower onto the cam surface, 2) the servo facility on this radial axis will be disabled during the digitising phase whilst the radius is measured via the 24 bit position registers on the integrated circuit.

REFERENCES

(1) Rank Taylor Hobson, Product Range Catalogue.

(2) McKeown, P.A. and Dinsdale, J. The Design and Development of a High Precision Computer Controlled Cam Grinding Machine. Annals of the CIRP, Vol. 23/1/74.

(3) Hannah, J. and Stephens, R.C. Mechanics of Machines (2nd Edition).

(4) HCTL-20xx Data Sheet. Hewlet Packard.

(5) HCTL-1000 Data Sheet. Hewlet Packard.

(6) ACT500 & S7AC Data Sheets. RDP Electronics.

(7) 3000 series Data Sheet. Hohner (GB).

(8) Turbo C v1.5 Reference Manuals. Borland International.

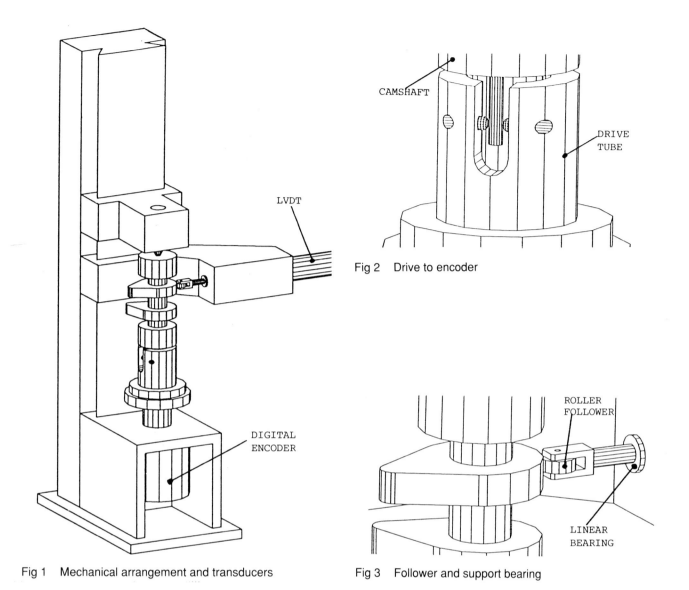

Fig 1 Mechanical arrangement and transducers

Fig 2 Drive to encoder

Fig 3 Follower and support bearing

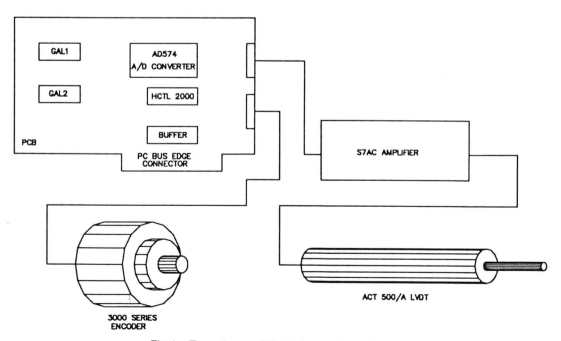

Fig 4 Transducer – PC interface schematic

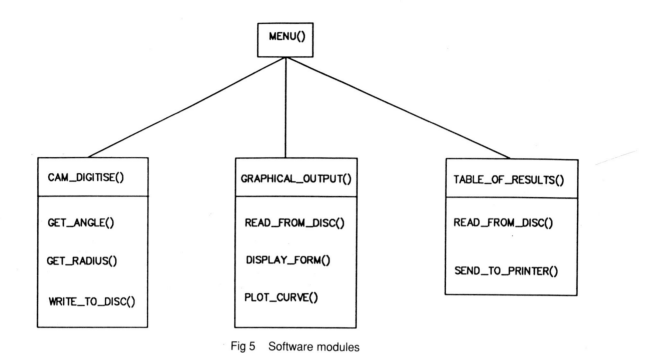

Fig 5   Software modules

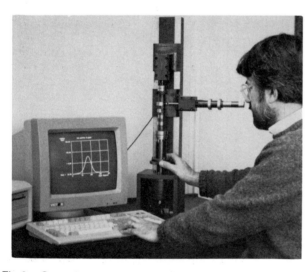

Fig 6   Operating the instrument

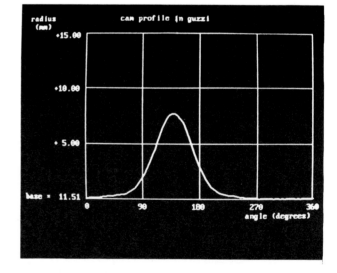

The appropriate control of manipulators for rehabilitation robotics

P J KYBERD, PhD, MSc, MISPO
Oxford Orthopaedic Engineering Centre, UK

SYNOPSIS

The choice of the type and form of operator control for a mechanical arm is important for the successful application of the device. It is necessary to choose the form that is appropriate to the user's aims, abilities and environment. When the arm is to be used for long periods, especially by people whose communication channels are limited by physical or mental barriers, the choice of input becomes critical. In addition, the design of the end effector can make the control of the system easier by mimicking human systems and so using the structures inherent in the human nervous system.

1 INTRODUCTION

Any new aid that is designed for frequent use by humans, must, by its very nature, be easier to use than the alternatives. No matter how impressive, attractive or expensive a device is, (especially a robotic aid) it will rapidly be rejected if it can be replaced by a far simpler method. While this is always prudent engineering practice, when robotics and electronic control is to be applied to rehabilitation it becomes essential to pay close attention to the abilities of the users.

When a human arm is engaged in the act of manipulation the various processes that control the action form a hierarchical structure, Figure 1. High level strategic commands dictate the overall performance. Subconscious layers prepare the hands orientation and approach, the hands form is dictated by a knowledge of the target object, its weight, shape and the act intended for it. Separate areas of hand posture and force control coordinate to frame the correct response. Finally, the digits' positions are controlled individually by internal feedback loops as well as feeding position and force data back up the chain of command [1].

To control a mechanical arm such as a robot for use by a disabled operator or an artificial arm prosthesis, it becomes an important question as to where the chain of command is broken. The choice can ensure adequate and easy control of the arm, or information overload for the operator.

1.1 Teleoperation

It is common to make the command insertion at level 1. This is conventional teleoperation. The break severs the feedback paths from the sensors to the digit control whilst burdening the operator with the detailed control of the individual joints down to the end effector. This requires great concentration on the part of the operator and rarely allows the user to function efficiently for long. For prostheses this would be burdensome. If any type of prosthesis is to succeed it must be easily fitted to the person and its operation quickly learned by them, otherwise few people will have the patience to persevere.

There have been attempts to rectify this break in the feedback paths by supplying forms of feedback to the operator at other sites on their body, for example, vibration or forces applied further up the arm [2,3,4]. The flaw in this is that it requires the operator to make a transformation from the data input to the parameter to be detected. Human beings are perfectly capable of achieving such feats (after training) but again it requires perseverance and motivation to succeed. For a robot in a rehabilitation set-up the user may have very few feedback paths open to them and such complexities would mitigate against their successful use of the robot.

This form of teleoperation is limited to use in controlled environments by highly trained and motivated individuals, such as the nuclear industry in remote handling situations.

1.2 Program control

Groups working in rehabilitation or conventional robotics apply the control from the highest level (4). This is function based robotics. The robot repeatedly performs a series of preprogrammed acts, commonly without any real external feedback. The majority of assembly line robots fall into this category. Only the most modern generations of robots have been given force sensitivity for manipulation.

Examples of this form in the rehabilitation field include the CEA Master [5] and the CURL controlled robots [6]. The control relies on being able to define the entire environment or to use vision systems to update the robots knowledge of the world. Even so that world model is a small one and limited in scope. These robots best operate in a workstation type environment, the robot performing the task chosen from a menu. For CURL (The Cambridge University Robot Language), the robot is driven by an object-orientated language that can control the robot to perform single action (such as "Grasp the cup"), Or strings of simple actions, ("Get the cup"). Or complete acts: "Drink", which would include getting the cup and bringing it to the mouth area and then withdrawing it and placing it down when the person signals they are finished, [5]. Full CURL is also updated by an overhead camera and its knowledge of the environment ensures that when objects are moved in the target area the program can detect this, also the objects are lifted over any other item in the field.

In addition, a level of user intervention to such systems has been added, the robot may follow pre-programmed tasks but the operator corrects the errors or changes in the layout of the target object just prior to manipulation.

Given these limitations the robots have been successful, but there is the occasional suggestion that a complex arm prosthesis might also perform in such a pre-fixed manner, completely ignoring the human operator. Such constraints would remove the spontaneity completely, for example having to always stand in a particular position relative to a target in order to pick it up. This is totally unacceptable.

1.3 Appropriate feedback

Earlier, difficulties with teleoperation were mentioned briefly. The biggest problem with some of the forms of feedback that are applied to teleoperation is that it may not appropriate to the task. When the contact force is fed back to the operators finger tips this is appropriate and easily learned. For example; A robot from Hitachi's research and development centre in Takasago [7,8]. However, unlike conventional teleoperation arms, the arm's disposition information is not made available to the operator, lessening the potential advantages gained.

Other forms of teleoperation are made virtually feedbackless using only a joy-stick which does not have any anthropomorphic form at all, only vision remains. Instead *if* the feedback control input is made at level 2 and *if* the feedback of the correct manner, success is possible, with very little direct training. A form of prosthesis control of this type is known as Extended Physiological Proprioception or EPP [9,10]. EPP was pioneered by D C Simpson at the Princess Margaret Rose Hospital in Edinburgh, as the control of whole arm prostheses for children born with the effects of the stereo-isomer of thalidomide.

The principle of EPP is simple: Force feedforward, position feedback. For example, a gripper controlled by EPP, Figure 2. A lever drives the gripper, ether via the rigid link, or by a servo loop. If the control level is pushed forward, it only results in movement of the lever once the robot gripper or prosthesis moves and sends this information back to the lever's servo, in this way the operator 'feels' the motion. This is easy to learn because the information is appropriate and correctly applied. It is appropriate as it is in the manner that human movement is controlled, (muscular tension gives rise to motion, which is what is felt by the person. The information is applied to the correct point because its motion feedback is returned to the commanding input. Thus the system forms an 'unbeatable servo' that can be tuned to the gain appropriate to the task. Thus a motion impaired operator who can only make small motions can still control the full range of movement of an arm or robot.

The flaw with EPP is that it controls end point motion of the arm. It still must have a simple manipulator on its termination or loose its advantage with additional manipulator controls. It cannot easily hold objects that are not designed to be held by it. In the unstructured environment of the home where a rehabilitation robot or prosthesis must operate, the household objects are normally designed to be held by the human hand, for EPP (or conventional prostheses) such items would have to be modified.

2.0 Human control

The hand is a complex device with many articulations which allow it to perform a very wide range of different tasks such as a tool, a weapon, a manipulator, an explorer and an expressor. As a manipulator it can form two major grip types: Precision, where the tips oppose (allowing delicate manipulation), and Power, when they wrap around the target object (providing a strong grip for forceful actions). Many grippers, particularly conventional prostheses cannot achieve both grips with the same terminator. A high level of precision achieved by human hands results from the ability of the thumb to effectively oppose any number of the hand's fingers from a two point grip or a three jaw chuck up to all the digits. For a prosthesis to achieve this it would require a more complex design than commercial examples.

One of the most advanced prostheses produced is the family of Southampton hand prostheses. In its most sophisticated form it possesses four degrees of freedom [11,12]. This allows the index finger to be flexed separately from the other three fingers, and thumb to oppose the tips or side of the fingers allowing the hand to adopt the major grip types in both two and three point manner as well as side opposition for carrying trays etc., (Figure 3). Indeed the hand can be held in a flat position this is not only cosmetically pleasing but allows it to be placed down a sleeve or in a pocket.

This control form places a break at level 3. The user makes a simple strategic decision and instructs the hand to hold the object. A micro-controller is then able to select a grip from the postures that is most appropriate to the task and dynamically adjusts the posture to maximise the contact area. Grip tension is thus minimised and it is controlled by the sensors on the hand detecting when the object slips and increasing the grip tension to compensate. The user is freed from the chore of concentrating upon the direct control of the hand and need only make strategic decisions. The hand is operated by the user commanding it with simple instructions to open, allowing the target object to be admitted. Then the hand can close on the object applying the lightest possible pressure. Force sensors on the palmar surface detect the objects shape and adapt the hand to it. Once the hand is commanded to hold the object sensors detect if the object is slipping and automatically increase the grip pressure to stop the slide. When used in rehabilitation robotics the advantages are clear: The disabled operator has few channels of input/output to control and the Southampton system minimises those it will dominate.

The controller and hand design was originally posed to overcome the short-fall in conventional powered prosthesis control which is mainly open loop. Visual cues are available but these can be limited by obstructions etc. The hands are simple single degree of freedom devices that cannot detect the applied tension. The commonest form has the hand fixed in a precision grip with the fingers in a three point chuck [13]. The thumb and fingers flex with equal angular speed. This design is flawed for two important reasons: Firstly, the range of objects that can be held are limited by the degree the hand can open or close. The standard designs cannot open wide enough to hold a pint glass nor close down sufficiently to hold the handle of a kettle in a power grip. In addition, when a small object is picked up from a horizontal surface the index finger and thumb must be brought close to the target. The other digits which are fixed in parallel with the index finger, obstruct this approach as they lie below the index finger. To overcome this problem users often invert the hand, palm facing outwards and the index finger closer to the surface. This places the hand in a posture that would rarely be adopted by someone with a natural hand.

The second problem with this design is the result of the subtle ways in which humans control their own fingers. When a person uses their hand to pick up an object they flex their fingers through a greater angle than their thumb. The thumb is brought close to the object and the fingers close on to it. This allows the thumb to be used for visual confirmation of the correct position of the hand, i.e. the number of degrees of motion are limited to simplify the task. In a similar way human babies when learning to grab objects limit the motion to a series of one degree of freedom swipes.

This use of the thumb was graphically demonstrated by Wing [14,15] using a two dimensional recording technique. Recently, this has been confirmed at the Oxford Orthopaedic Engineering Centre by the author using a three dimensional motion recorder that allows a more natural grip attitudes to be adopted. The original experiment used a single overhead camera and vertically orientated objects. The 'Vicon' System automatically captures the motion of an object in space using two or more video cameras and retroflective markers on the landmarks of the moving object and then reconstruct it into motion in three dimensions to a resolution of 0.5mm or better.

A typical reaching motion is shown here, the motion is relative to an object of 40mm diameter, (Figures 4 and 5). The motion of wrist and the tips of the index finger and thumb are indicated. As the hand closes on the object the index finger passes through a phase where it is held further from the target object until the thumb is closer in. Then the thumb is held as a strut from which the fingers closure can be guided. So strong is this tendency that when the angular velocity of the digits is equal (as in a prosthetic hand) the person may correct for this. Wing showed that for a single individual used shoulder motion to counteract the thumb's motion of her artificial hand. Whilst this is a little awkward physically, it ensured she was using the same 'targeting algorithm' for both her natural hand and her artificial hand.

The alternative form of prosthesis control is the cable driven, (body powered) hook or hand. This provides the operator with some degree of feedback on the status of the hand but it is limited to a single degree of freedom and requires a range of specialist terminators to be used in addition. If the most functionally adaptable terminator is used, (the split hook), many normal jobs can be performed, however, the device cannot be described as attractive. It was this gap that the Southampton hand was designed to fill. The hand is anthropomorphic and so it can be covered with a cosmetic glove and can be passed off as a normal hand in social situations. The additional degrees of freedom provide the adaptable range of the hook, however the control must be then simple for the hand to be accepted widely, thus the Southampton Anthropomorphic Manipulation Scheme, SAMS (as outlined earlier) is used. A small number of individuals have used this scheme to control prostheses with success, finding it easy to learn and use. The tests used a range of activities designed by the DHSS to assess artificial limb users. The SAMS hand performed much better than a conventional electric hand and on a par with the split hook. While this is encouraging it is worth noting that the subject (while being exceptionally bright) was exclusively a hook user in every day life [16,17].

To expand the hand beyond the laboratory a simplified SAMS hand was designed based on the conventional electric hand. The device was modified to take a custom designed touch and slip sensor and a single board 8 bit micro-controller system was implemented to control the hand [18]. Two users of electric hands were introduced to the device. The first was in the laboratory. The second was trained and an arm to fit him was manufactured to take the hand. This was briefly used at home and at work until financial constraints curtailed the tests. However, they both found the control easy to learn and use especially once they had unlearned the manner in which they controlled their conventional arms, which involves higher levels of concentration [11,13,18].

The most common way that electric hands (including the Southampton hand) is instructed is using the voltages generated when muscles contract (electromyograms or EMGs). These are filtered, amplified, smoothed and used as a voltage input to the controller. Naturally the origin of the command voltage need not be a muscle, any other source will suffice after appropriate interfacing. With such an adaptable system as a computer based controller it can be used in a wider range than outlined previously. For example, the hand is equally exploitable as robot manipulator. While the anthropomorphic appearance is unnecessary the shapes it can adopt which are appropriate to holding objects designed to be held by humans is useful [20].

The Oxford Robot Assistant Controller (ORAC), is a robot test vehicle designed to be used by individuals with motion impairment to perform simple every day tasks [22]. The group of individuals who will benefit from robots for rehabilitation and those with substantial impairment in their own movement or control. An important group of these people are those with degenerative diseases of their motor control, (Motor Neurone Disease or Muscular Dystrophy). These disorders result in the loss of precision or power in the muscles from the degeneration of the neurones to their muscles. The individuals are often struck with the encroaching loss once they are middle aged. Their cognitive function is completely intact, but they are tragically trapped in a barely functioning body. For this application it becomes possible to leave the high level intelligence of the system to the user, only the detailed control need be maintained by the computer system. The robot is mobile but the navigation is performed by the user.

The robot is teleoperated using EPP with an automatic gripper based on the SAMS technology. This ensures that the limited number of inputs the operator has are not occupied with instructing the hand the operator can issue the single command to hold.

In essence the computer system is designed to augment the operators abilities. For example the robot base would follow tracks laid down in the house whilst collision avoidance and direction are controlled by the operator. The robot follows a track until a choice of direction is possible, then the user guides it left or right depending on desire. Once parked at a workstation, or beside the wheelchair the robot will assist in feeding, preparing food, selecting reading material etcetera. The display screen is designed to make the selection of tasks easier, the mouse the robot and the base all are controlled by the same joy-stick.

A second group that can benefit from robotic devices is children with Cerebral Palsy, the control system replacing their own impaired Central Nervous System. The CURL robots from Cambridge have been used in a teaching environment when the children select from a menu that instruct the robot to perform cooking tasks.

The motivational part of this robot cannot be ignored. In an intelligent wheelchair system from the CALL centre in Edinburgh [6], and more recently with Keele's Handy 1 feeder robot, [5]. Children involved have been motivated, because they could (for the first time) control an aspect of their environment rather than having everything done for them. In the case of the Handy 1 it acted as a scoop feeder, allowing a child to develop targeting skills and control of his lips, tongue and swallowing reflex to such an extent he has moved beyond the Handy and can cope with only a stick and strap to help him feed.

3 Conclusion

The correct application of ergonomics to the design of equipment should be a watchword for all engineers. It is clear that this is not always the case, for example; the slow adoption of a better interface to computers than a typewriter keyboard, or *any* interface at all, onto domestic video equipment, [21]. In these examples the interface was instrumental in maintaining the slow pace of acceptance of

the devices in the past and the current under exploitation of their capabilities. While they were aimed at a small population of highly trained, highly motivated professional operators, these failures are obscured by the users skill and experience. Once the public had access to the equipment the failings became apparent.

In the field of rehabilitation where the operators physical abilities mean the input of a single letter or command takes considerable effort and concentration, any additional barriers, no matter how small, due to the poor design are likely to be completely insurmountable. The application of ergonomics therefore becomes paramount, Unless, or until, the computing/electronics technology improves to an extent where robots becomes cheap autonomous intelligences we must design systems that augment peoples' abilities and fully utilise their abilities. The best option is to use the controls and structures already in place in a human Central Nervous System, it is the selection of the correct type of information transfer is the central issue in this process.

References

1. JENNEROD M. The timing of natural prehension movements. *Journal of motor behaviour,* 16, 235 - 254.

2. PATTERSON P.E. and KATZ J. A. Design and evaluation of a sensory feedback system that provides grasping pressure in a myoelectric hand. *Journal of rehabilitation research and development.* 1989, 1, 1 - 8.

3. COGHILL-SMITH H. A. Data transfer via the skin. *International Society of Prosthetics and Orthotics,* UK scientific meeting, Nottingham, 12 -14 April 1989.

4. SOLOMONOW M., LYNN J. L. and FREEDY A. Fundamentals of electrotactile stimulation for sensory feedback from limb prostheses. *Sixth International symposium on the external control of human extremities.* Dubrovnik, Yugoslavia 1978.

5. JACKSON R., GOSINE R, MAHONEY R. M. (editors) *Proceedings of 2nd Cambridge workshop on Rehabilitation robotics,* Selwyn College, Cambridge, 12 - 14 April 1991.

6. HARWIN W. S., JACKSON R., GOSINE R, (editors) *Proceedings of 1st Cambridge workshop on Rehabilitation robotics,* Selwyn College, Cambridge, 14 -16 April 1989.

7. OOMICI T. et al, Development of working multifinger hand manipulator. *IEE International workshop on Intelligent robots and systems. IROS '90.* 873 - 880.

8. OOMICI T. et al, Development of an advanced robot manipulator system. *SMiRT 11 Transactions* SD1, August 1991. 343 - 349.

9. SIMPSON D. C. The control and supply of multimovement externally powered upper limb prosthesis. *The fourth international symposium on advances in the external control of human extremities.* Dubrovnik, Yugoslavia, 1972, 247 - 254.

10. GOW D. J., DICK T. D. et al. The physiologically appropriate control of an electrically powered hand prosthesis. *International Society of Prosthetics and Orthotics,* Fourth world congress., London, September 1983.

11. KYBERD P. J. Algorithmic control of a multifunction hand prosthesis. Phd thesis, 1990, Electrical engineering department, Southampton University.

12. BAITS J. C., TODD R. W. and NIGHTINGALE J. M. The feasibility of an adaptive control scheme for artificial prehension. *Institution of Mechanical Engineers, Proceedings 1968 - 69.* 183, 3J. Basic Problems of prehension, movement and control of artificial limbs.

13. ATKINS D. J. and MEIER R. H. Comprehensive management of the upper-limb amptuee. *Springer-Verlag*, 1989, ISBN 0-387-96779-6.

14. WING A. M. and FRASER C. The contribution of the thumb to reaching movements. *Quarterly journal of experimental psychology.* 35A, 1987, 297 - 300.

15. WING A. M. and FRASER C. Artificial hand use in grasp and release phase of reaching. *International symposium of teleoperation and control.* Bristol, July 1988.

16. CODD R. D. Development and evaluation of adaptive control for a hand prosthesis. Phd thesis, 1975, Electrical engineering department, Southampton University.

17. MOORE D. Development of a multifunctional adaptive hand prosthesis. Phd thesis, 1980, Electrical engineering department, Southampton University.

18. CHAPPELL P. H., KYBERD P. J., NIGHTINGALE J. M. and BARKHORDAR, M Control of a single degree of freedom artificial hand. *Journal of biomedical engineering,*9, July 1987, 273 - 277.

19. KYBERD P. J., MUSTAPHA N., and CARNEGIE F. CHAPPELL P. H. Clinical experience with a hierarchically controlled myoelectric hand prosthesis with vibro-tactile feedback. *In preparation .*

20. KYBERD P. J., and CHAPPELL P. H. The Southampton hand: A hierarchical approach to controlling prostheses. *2nd International workshop on rehabilitation robotics.* Newcastle on Tyne, UK. 5 - 7 September 1989.

21. THIMBLEBY H. Can anyone work the video? *New Scientist,* 23rd February 1991, 48 - 51.

22. TURNER-SMITH A. R. Annual report of the Oxford Orthopaedic Engineering Centre December 1991. *Oxford Orthopaedic Engineering Centre.* ISBN 0263-2535. 39 - 43.

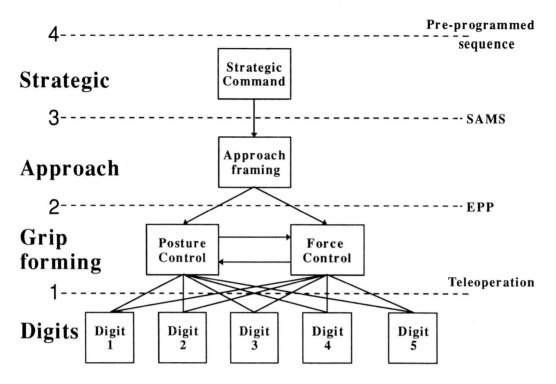

Fig 1 Schematic of the heirarchical structure of the control of limbs. The selection of the level of user input dictates the level involvement of the operator

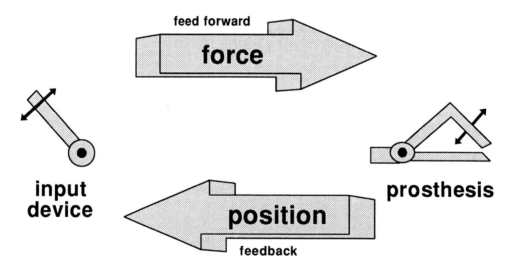

Fig 2 Extended Physiological Propriception control of artificial limbs. The control is force feedforwarded and position feedback

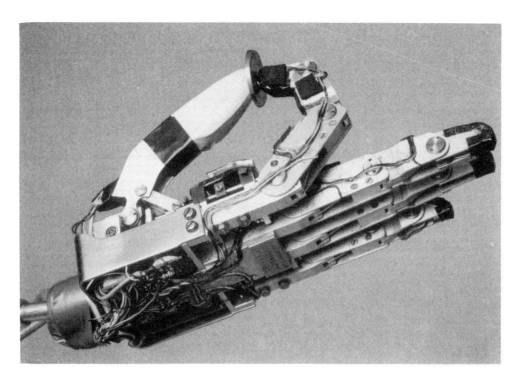

Fig 3   The Southampton Adaptive Hand Prosthesis, shown here in a two way, precision grip. This posture can be used to pick small objectives up from horizontal surfaces

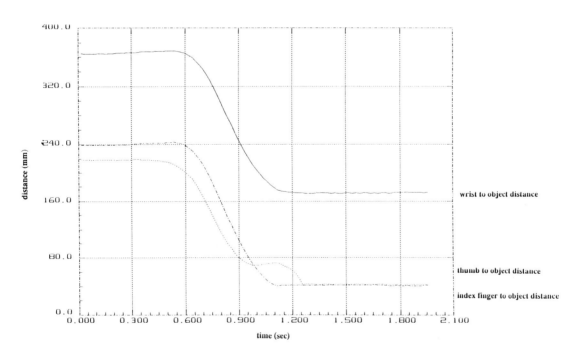

Fig 4   Approach of hand to target object. Distance of wrist, index finger tip, and thumb tip to object 40mm in diameter

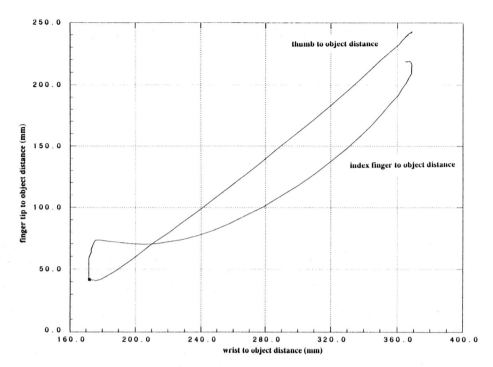

Fig 5 Tip to object distance against wrist to target distance. The thumb is brought in to the object and held as a strut relative to the target, the index finger is then closed upon it

Model reference adaptive control of a modular robot

P C MULDERS, A P M A MARTENS, J JANSEN
Technical University Eindhoven, The Netherlands

SYNOPSIS. Flexible manipulators often have place - and time dependent parameters, varying during trajectory performance. Adaptive control is a process of modifying one or more parameters of the controller of a manipulator in action and so specially important for these robots.
Here an adaptive controller is described as a combination of the computed torque method and an adaptive PD controller based on the Model Reference Adaptive Control (MRAC) method.
It has been applied to a modular robot - for loads up to 50 kg - consisiting of a linear and a rotary actuator showing these parameter variations.

Necessary models - extended and reduced - of this RT robot have been made and the proposed controller has been tested in simulations and in the real configuration also with respect to stability, convergency and robustness.

1. INTRODUCTION

This work is a study of (optimal) adaptive control algorithms on systems with place- and time dependent parameters - varying during trajectory performance - with an implementation on mechanical manipulators of industrial scale.
These advanced control systems are tested on a modular robot system - for loads up to 50 kg, consisting of a linear and a rotary actuator - as shown in Fig. 1.
A 3D-force sensor is mounted on the linear arm to perform trajectory teach operations. After that the replay of the desired trajectory - eventually with varying parameters - is done, in which the known control signals are updated by the adaptive control algorithm.

Fig 1   The modular robot system

Robotcontrol has been studied mostly under the assumption that actuators are stiff and that the links can be modelled as rigid bodies. Therefore most robots have a very stiff construction to avoid deformations and vibrations.
For higher operating speeds robots should be light weight constructions to enable the robots to respond faster.
Hence, more accurate dynamic models should be taken into account to pursue better dynamic performance.

Now a number of (optimal adaptive) trajectory control strategies may be mentioned here:

- the PID method
- the optimal controller (Riccati equation) method
- the model reference adaptive control (MRAC)method

The PID controller uses the deviation from the desired trajectory to correct and is often used as a standard to compare with other controllers. However for coupled systems this type leads often to instability.
The linear optimal controller is based on the minimization of a performance criterion function with contributions of e.g. the deviations and the control signals with certain weighing matrices.
Another approach to improve the behaviour of robots is the computed torque control method, sometimes called the inverse dynamics control. The necessary torques are calculated from the prescribed trajectory and so the control law is designed explicitly on the basis of a model. If flexibilities play a role, it often results in an unstable system behaviour.

So the aim is to search for a control law achieving both trajectory tracking and a stabilization of acceptable vibrations.
Adaptive control is the process of modifying one or more parameters of the structure of the control system to force the response of the closed loop system towards a desired trajectory.
Among these adaptive methods the model reference adaptive control method (MRAC) is important.
So the adaptive controller described here is a combination of the computed torque method as feed forward control, and a PD feedback controller acting on the deviation, while this PD controller is updated by the MRAC algorithm.
This leads to relatively easy to implement systems, with a high speed of adaptation and may be used in a variety of

applications.

From the modular RT robot system an extended model - for simulation purposes - has been made, while a reduced model is applied to calculate the computed torque signals. The described method has been performed both in computer simulations and reality to draw conclusions with respect to convergency, stability and robustness.

2. THE ROTATION - TRANSLATION ROBOT

2.1 The construction of the linear robotarm.

The linear arm consists of a hollow frame with a preloaded spindle. The rotation of the DC motor is converted into a translation of the arm with a ballscrew nut on the spindle. For the position - and velocity measurements there are a linear - and rotational encoder and a tachogenerator. The 3D-force sensor has strain gauges and is used in the teach-mode.

velocity	: 1 m/s
acceleration	: 1 m/s^2
load	: 50 kg
stroke	: 1 m
accuracy	: 0,01 mm
position measuring	: Heidenhain LS513
motor	: Axem MC19PR26, 1 kW
control system	: PID-or state controller.

Table 1. Linear robotarm specifications.

2.2 The construction of the rotational module.

The mechanical construction is based on a cylinder with side ribs to minimize the deformation. The transmission from the motor to the turntable consists of a four stage preloaded toothed wheel combination. Coupled to the DC motor there is a tachogenerator and a rotational encoder. Direct position measurement of the turntable occurs by a bended optical digital incremental encoder.

velocity	: $\pi/2$ rad/s
acceleration	: $\pi/2$ rad/s^2
range	: $\pi \to 2\pi$ rad
accuracy	: 10^{-5} rad
positon measuring	: Heidenhain LIDA 360
motor	: BBC - MC 19P, 1kW
control system	: PID-or state controller

Table 2. Rotational module specifications.

2.3 The hierarchical controller.

The controller (Fig.2) consists of 4 Intel SBC's and 1 RAM board.

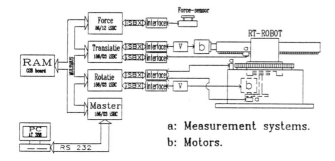

a: Measurement systems.
b: Motors.

Fig 2 The hierarchical controller

The task of each slave SBC 186/03 is:
  to calculate - according to the control algorithm - the motor voltages
. to read the position of each module
. to store these data - motor voltages and positions - into the RAM.

The task of the master SBC 186/03 is:
. to synchronize the software in both the other SBC's.
. to transfer the data over the RS 232 bus.

The PC 80386 may calculate the optimal control law and the nominal trajectory off-line and diagram the data.

2.4 Modelling of the RT-robot.

Although the robot is a system with divided parameters an attempt is made to realize a lumped mass model. This approach is based on previous studies [1] about drives of motor-tacho-spindle-carriage combinations. The extended model has 11 degrees of freedom (DOF):

$$q = [\varphi_0 \; \varphi_1 \; \varphi_3 \; \varphi_5 \; \varphi_7 \; \gamma \; \psi_0 \; \psi_1 \; \psi_2 \; s_1 \; x]^T$$

By the coupling of the modules for rotation and translation also the eigenfrequencies may vary. So the lowest eigenfrequency of the rotational module is $f_o \approx 18...20$ [Hz] and of the linear actuator is $f_o \approx 110 ... 134$ [Hz].

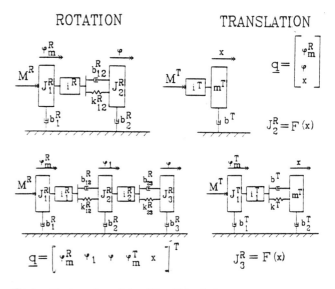

Fig 3   Reduced models of the RT-robot

The complexity of the simulation model may be reduced as a compromise between accuracy and duration of simulation calculations. The lowest eigenfrequency should be present in the reduced model. But there is another reason for model reduction i.e. the realization of a controller (e.g computed torque) via the control model.
With the combination of a simulation model (5 DOF) and a controller based on the control model (3 DOF) rather good results have been obtained. This controller has been implemented with good result in the RT-module.

3. MODEL REFERENCE ADAPTIVE CONTROL

For the RT-robot a non adaptive and an adaptive control have been designed. (Fig. 4). In the non-adaptive case the adaptation algorithm is out of operation, so these may be well compared.

Non-adaptive controllers require exact knowledge of the systemparameters and explicit use of the complex system dynamics. In practice one has to deal with uncertainties - so a number of parameters as moments of inertia, loads and armlength may vary, while non- linearities may be unknown-leading to a bad performance of the controller.
The application of feedback may reduce the sensitivity for parameter variations, but this leads to higher gain factors, bigger control efforts and a possible instability.

In adaptive control the model parameters of the system are estimated on-line. Based on this estimation the control effort is determined. So adaptive control is very suitable for manipulators, with a complex system description with unknown and varying parameters.

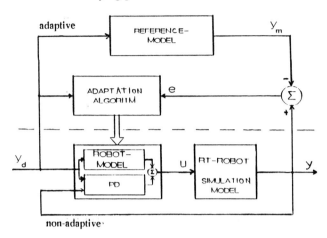

Fig 4   The non-adaptive and adaptive control concept

In this chapter an adaptive controller is proposed, which is a combination of the computed torque method for the main control input and an adaptive PD controller acting on the deviation of the desired trajectory.
The computed torque signal is derived directly from the equations of a <u>control model</u>.
An adaptation algorithm based on the Model Reference Adaptive Control (MRAC) method adapts the PD-gain factors on line.
The complete controller is applied to a <u>simulation model</u> as well as the real RT-robot.

3.1 The non-adaptive controller.

The nominal control efforts - the torques - to perform a desired trajectory are computed by a <u>control model.</u> This has to be a representative reduced model and not too complex, otherwise the computation time of this part of the control signal becomes too big. The 3 DOF-model (R2T1) has been applied here.
Next the PD-controller acts on and compensates for the realized trajectory error.

So the control signal consists of: $\bar{u} = \bar{u}_{model} + \bar{u}_{PD}$ (1)

The control model has three degrees of freedom:

Rotation of the rotation motor : ϕ_m^r
Rotation of the turntable : ϕ
Translation of the linear arm : X

The non-linear equations of movement are:

Rotation:

$$M^R = J_1^R \ddot{\phi}_m^R + \left[\frac{b^R}{(i^R)^2} + b_1^R\right]\dot{\phi}_m^R - \frac{b^R}{i^R}\dot{\phi} + \frac{k^R}{(i^R)^2}\phi_m^R - \frac{k^R}{i^R}\phi \quad (2)$$

$$0 = J_2^R \ddot{\phi} + \frac{\partial J_2^R}{\partial t}\dot{\phi} + \left[b^R + b_2^R\right]\dot{\phi} - \frac{b^R}{i^R}\dot{\phi}_m^R + k^R\phi - \frac{k^R}{i^R}\phi_m^R \quad (3)$$

Translation:

$$M^T = \frac{1}{i^T}\left[m_T \ddot{x} - \frac{1}{2}\frac{\partial J_2^R}{\partial x}\dot{\phi}^2 + b^T \dot{x}\right] \quad (4)$$

The desired trajectory - with estimates of the other degrees of freedom - is substituted in equations (2) to (4) to calculate the nominal control torques. So this part of the input signal is:

$$\vec{u}_{model} = \left[M^R(q_d, \dot{q}_d, \ddot{q}_d) \quad M^T(q_d, \dot{q}_d, \ddot{q}_d)\right]^T \quad (5)$$

The real trajectory is compared with the desired trajectory and so the PD control effort is obtained:

$$\vec{u}_{PD} = -K_d \vec{e} - K_p \dot{\vec{e}} \quad (6)$$

K_p and K_d are of the following structure according to the assumption that deviations in the rotation or translation only lead to a control effort in that degree of freedom.

$$K_p = \begin{bmatrix} K_4^R & K_1^R & 0 \\ 0 & 0 & K_1^T \end{bmatrix} \quad K_d = \begin{bmatrix} K_5^R & K_2^R & 0 \\ 0 & 0 & K_2^T \end{bmatrix} \quad (7)$$

The feedback gains are determined such that the total system is stable with poles in the left half of the s-plane.

3.2 The adaptive controller.

Adaptive control is a kind of feedback, in which the states of a process are divided in two categories, characterized by the difference in speed. It is assumed that the model parameters are slowly changing, while the degrees of freedom are quickly changing states. (Fig. 4).
The fast control loop is the computed torque part and the PD-controller. The systemparameters and subsequently the control parameters (modelparameters and feedback gains) are not constant, but they are updated in a slower control loop as an answer to the change in dynamics of the process and to disturbances.
In the slow control loop there is a reference model which describes the desired trajectory in terms of the deviation.
The control parameters are determined such that the robot is forced to behave as the reference model. The adaption mechanism estimates on line the control model parameters and feedback gains by using the deviation and the reference model.

3.3 The adaptation algorithm.

The adaptation of the control parameters is done here by the model reference Adaptive Control approach given by Seraji [2].
The control effort consists again of a computed torque - part (control model) and a PD - part as given in (1).
The model feedforward part may be written as:

$$M^R = A^R \ddot{\phi}_m^R + B_1^R \dot{\phi}_m^R + B_2^R \dot{\phi} + C_1^R \phi_m^R + C_2^R \phi + F^R \quad (8)$$

$$M^T = A^T \ddot{x} + B_1^T \dot{\phi} + B_2^T \dot{x} + C_2^T x + F^T \quad (9)$$

so: $\vec{u}_{model} = A(t)\ddot{\vec{q}}(t) + B(t)\dot{\vec{q}}(t) + C(t)\vec{q}(t) + \vec{F}$ (10)

with: $A(t) = \begin{bmatrix} A^R & 0 & 0 \\ 0 & 0 & A^T \end{bmatrix} \quad B(t) = \begin{bmatrix} B_1^R & B_2^R & 0 \\ 0 & B_1^T & B_2^T \end{bmatrix}$

$$C(t) = \begin{bmatrix} C_1^R & C_2^R & 0 \\ 0 & 0 & C^T \end{bmatrix} \quad F(t) = \begin{bmatrix} F^R \\ F^T \end{bmatrix} \quad (11)$$

The PD-part is: $\vec{u}_{PD}(t) = -K_d(t)\vec{e}(t) - K_p(t)\dot{\vec{e}}(t)$ (12)

with:

$$K_p = \begin{bmatrix} K_1^R & K_2^R & 0 \\ 0 & 0 & K_1^T \end{bmatrix} \quad K_d = \begin{bmatrix} K_3^R & K_4^R & 0 \\ 0 & 0 & K_2^T \end{bmatrix} \quad (13)$$

This results in totally in 17 controlparameters i.e. 11 in the model-part (11) and 6 in the PD-part (13).
The total control effort, as the sum of (10) and (12) becomes:

$$\vec{u}(t) = -K_p(t)\vec{e}(t) - K_d(t)\dot{\vec{e}}(t) + A(t)\ddot{\vec{q}}(t) + B(t)\dot{\vec{q}}(t) + C(t)\vec{q}(t) + \vec{F} \quad (14)$$

The behaviour of the robot is described by a non-linear equation with unknown parameters (A^*, B^*, C^*, and $\vec{F}^*$):

$$\vec{u}_{robot}(t) = A^*(\vec{q},\dot{\vec{q}})\ddot{\vec{q}}(t) + B^*(\vec{q},\dot{\vec{q}})\dot{\vec{q}}(t) + C^*(\vec{q},\dot{\vec{q}})\vec{q}(t) + \vec{F}^* \quad (15)$$

Combining (14) and (15) results in:

$$A^*\ddot{\vec{e}}(t) + (B^* + K_d)\dot{\vec{e}}(t) + (C^* + K_p)\vec{e}(t) = (\vec{F}^* - \vec{F}) + (A^* - A)\ddot{\vec{q}}_d(t) + (B^* - B)\dot{\vec{q}}_d(t) + (C^* - C)\vec{q}_d(t) \quad (16)$$

The deviation will asymptotically not become zero, but depend on q_d and F. Therefore A,B,C and F have to be adapted such that the right hand side of (16) becomes zero.

The feedback gains K_p and K_d are also adapted to get stability of the closed loop at the desired performance.

After defining the position - velocity error $\vec{z}(t)$ eq. (16) is transformed into an adaptive system.

In the reference model the desired trajectory is described in the error $\vec{\varepsilon}(t)$ and it is assumed that the error of each DOF is decoupled and described as a second order differential equation.

$$\vec{z}(t) = (\vec{\varepsilon}(t), \dot{\vec{\varepsilon}}(t))^T \quad (17)$$

$$\ddot{\varepsilon}_i(t) + 2\xi_i\omega_i\dot{\varepsilon}_i(t) + \omega_i^2\varepsilon_i(t) = \vec{0}; \quad i = 1,2,3 \quad (18)$$

With ξ_i the relative dampingsfactor and ω_i the undamped eigenfrequency and $D_1 = \text{diag}(\omega_i^2)$ and $D_2 = \text{diag}(2\xi\omega_i)$ as constant 3 x 3 diagonal matrices it follows:

$$\ddot{\vec{e}}_m(t) + D_2\dot{\vec{e}}_m(t) + D_1\vec{e}_m(t) = \vec{0} \quad (19)$$

$$\dot{\vec{z}}_m(t) = \begin{pmatrix} 0 & I \\ -D_1 & -D_2 \end{pmatrix} \vec{z}_m(t)$$

The reference model is stable, so there exist a symmetric positive definite 6 x 6 matrix P, wich obeys the Lyapunov equation in which D is a 6 x 6 system - and Q is a symmetric constant 6 x 6 matrix.

$$P = \begin{pmatrix} P_1 & P_2^T \\ P_2 & P_3 \end{pmatrix} \qquad PD + D^T P = -Q \quad (20)$$

From this the adaptation algorithms are derived so that for a trajectory the state of the adaptive system converges to the reference model. The unknown robotparameters A^*, B^*, C^* and F^* are slowly time dependent compared with the adaptation.

For the control - and systemparameters follows e.g.

with $R(t) = P_2 \vec{e}(t) + P_3 \dot{\vec{e}}(t)$ \quad (21)

$$\dot{\vec{F}}(t) = \delta \vec{R}(t)$$
$$\dot{K}_p(t) = \alpha \vec{R}(t) \vec{e}^T(t)$$
$$\dot{K}_v(t) = \alpha_1 \vec{R}(t) \dot{\vec{e}}^T(t)$$
$$\dot{C}(t) = \beta \vec{R}(t) \vec{q}_d^T(t)$$
$$\dot{B}(t) = \gamma \vec{R}(t) \dot{\vec{q}}_d^T(t)$$
$$\dot{A}(t) = \lambda \vec{R}(t) \ddot{\vec{q}}_d^T(t)$$

So $\quad \vec{F}(t) = \vec{F}(0) + \delta \int_0^t \vec{R}(t) dt \quad$ etc.

So summarizing the main properties of the adaptive control concept are:
1. Two control loops, a fast loop for the degrees of freedom and a slow loop to adapt the control parameters.
2. The control parameters are adapted on-line.
3. Feedback takes place from the performance of the fast loop.

4. RESULTS

4.1 Simulations.

The simulations have been performed with the package PC-Matlab. The simulation-model of the robot has 5 DOF ($R_3 T_2$), while the control model for the computed torque part has 3 DOF ($R_2 T_1$). The desired trajectory is a skew sine wave in both φ and x

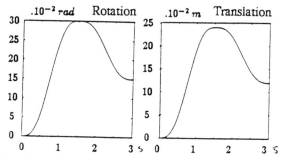

Fig 5 The desired trajectory

The minimal sample time is 7 ms, applied in the simulations and the implementation.
In Fig. 6 the results of the non-adaptive controller are shown with and without the computed torque part (feed forward).
Feed forward control improves the control performance considerable.

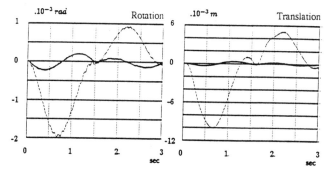

Fig 6 Position errors with the non-adaptive controller
(_____ with feedforward)

A comparison between the performances of the adaptive - and the non-adaptive controller (with a load of 0 kg) is

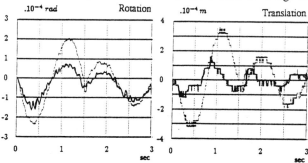

Fig 7 Performance adaptive vs non adaptive controller
(........ non adaptive)

shown in Fig. 7. The initial conditions of the control parameters are at the start of the trajectory the same for both the adaptive and the non-adaptive controller. The adaptative controller performs better than the non-adaptive controller because of the adaptation mechanism.

The performance of the adaptive controller on different loads is shown in Fig. 8. The position errors are reduced by a factor 2 for rotation and 6 for translation compared with the non-adaptive controller.

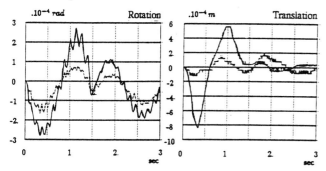

Fig 8   Position errors-adaptive controller-different loads
        (........ 0 kg; ____ 50 kg)

4.2 Implementation on the RT-robot.

The same experiments as in the simulations have been performed with the real RT-robot. The adaptive controller needs a sufficient long trajectory to estimate the control parameters well, so here the trajectory consists of four skew sine waves.
In Fig. 9. a comparison is made between the adaptive and the non-adaptive controller.
The RT-robot is rather stiff, so small variations in the load are easily compensated by the PD-controller. With a load of 20 kg the adaptive controller tends to perform better.
If the control model is chosen such that the parameter values are 30% lower than the real RT-configuration, then the adaptation mechanism updates the control parameters such that the control performance rather quickly becomes better, as shown in Fig. 9, which means a good robustness.

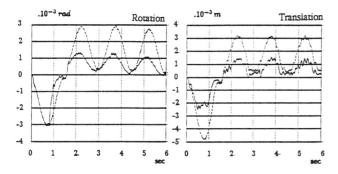

Fig 9   Position errors-adaptive and non-adaptive
        controller RT robot (........ non adaptive)

4.3 Robustness and Adaptation Speed.

In the case of different loads the control performance becomes less but no instabilities occur. Also if the feedback gains are not chosen properly, then the response of the real robot may become unstable. The adaptive controller however will try to stabilize this effect. This is called the robustness.
The adaptation mechanism is able to stabilize an initial unstable controller. It also restricts the feedback gains to become negative.
If the controlparameters are updated only every 20 samples then the occuring errors are hardly different from the fast updating situation.

CONCLUSIONS

The application of feedforward (computed torque) control derived via a control model from the desired trajectory improves the control performance considerably.
The experiments have been done with maximum adaptation speed. Speed reduction by a factor twenty, gives no significant difference in the position error.
The non-adaptive controller is sensitive to load variations, so a load of 50 kg makes the control performance worse.
The adaptive controller is preferable if the robotdynamics are poorly known. In that case the non-adaptive controller will give a bad control performance and possibly lead to instability.
The adaptation mechanism estimates the best control parameters and is an improvement compared to the non-adaptive controller. The adaptive controller is also rather robust. An initial deviation of parameter values of the control model with 30% causes the adaptation mechanism to update the controlparameters quickly and results again in a good control performance.

LITERATURE

(1) Mulders P.C. Oosterling J.A.J. Wolf A.C. H. van der.
    A model study of a feeddrive for a numerical controlled lathe.
    CIRP-Annals 1982, Vol 31/1, pp 293-298

(2) Seraji H., Design of adaptive joint controllers for robots.
    Recent Trends in Robotics pp 251-260 Elsevier, Amsterdam 1986.

(3) Seraji H., Adaptive Control of Robot manipulators.
    Proc. IEEE, Conf.on Robotics and Automation,
    San Francisco 1986.

(4) Voorkamp R.J., Study of the adaptive trajectory control of a RT-robot, WPA-1076, MS-thesis, Technical University Eindhoven, 1991.

(5) Mulders P.C., Jansen J., Pijls J.M.L., Optimal trajectory control of a linear robotarm by a state-space method.
    CIRP-Annals 1989, Vol 38/1 pp 359-364.

(6) Kreffer G.J. Design of a rotational module and a study of adaptive trajectory control of the RT-robot.
    WPA 0575 MS thesis
    Technical University Eindhoven. 1988.

(7) Bax W.H.M Study of the dynamic model of a RT-robot   WPA 0787 MS thesis
    Technical University Eindhoven. 1989.

(8) Martens A.P.M.A Comparative study and implementation of a trajectory control of a RT-robot.
    WPA 0955 MS thesis
    Technical University Eindhoven. 1990.

Digital control of pump-actuated hydraulic manipulators

J E HOLT, BSc, AMIMechE, **D E B PALMER**, MA
Nuclear Electric plc, UK

SYNOPSIS

Nuclear Electric has for many years made use of fixed base telemanipulators for carrying out inspection and minor refurbishment work inside its reactors. In many cases hydraulic actuation has been selected for its high power to volume ratio. Manipulator joints are operated by separate pump units, one per joint. In order to meet ever more demanding applications, ways have been sought of using digital algorithms to improve joint control. A theme has been the improvement by digital means of existing types of actuator, obviating the need for extensive redesign of mechanical and hydraulic systems. This paper will examine two types of pump unit and two applications. A digital controller system capable of controlling either unit and implementing a wide range of algorithms is described.

1. BACKGROUND

Nuclear Electric PLC and its predecessor, the CEGB, have for many years made use of remote controlled inspection and maintenance equipment inside their reactors. In particular the need to extend the life of the older generation Magnox plant, whilst maintaining a very high and increasing level of safety assurance, has led to the need to carry out remote operations never envisaged when the stations were built. A common method of carrying out work in areas where no man access can be allowed is the use of fixed base telemanipulators. Such a machine typically gains access to the interior of a reactor during a shut-down via a refuelling standpipe. The work is mostly non-repetitive, and there is a need to exercise great care when working on the plant. Thus manipulators are designed for slow, precise movement and are always controlled or supervised by a human operator. There is little or no scope for automation of the type seen with production robots.

In many cases, hydraulic actuation has been chosen, since this provides by far the greatest load carrying capacity. Manipulator dimensions are limited by the need to pass through a fuelling standpipe approximately 200 mm in diameter. Figure 1 shows a typical hydraulic manipulator in position. Multi-link manipulators of this type are typically used for TV inspection purposes, and for the carrying and positioning of work packages capable of carrying out simple repair operations. They are generally operated using one joystick for each joint, the joystick controlling the angular velocity of the joint. Whilst this method of operation is quite suitable for many uses, the relatively large oil volumes, the long and partly compliant hydraulic lines, and the gravitational and friction loads all limit the responsiveness of the system, making fine control difficult. Furthermore, it is beyond the operator's ability to make any sort of coordinated simultaneous movement of several joints, for example to avoid an obstacle or to move the manipulator tip in a straight line. Use of a computer guidance

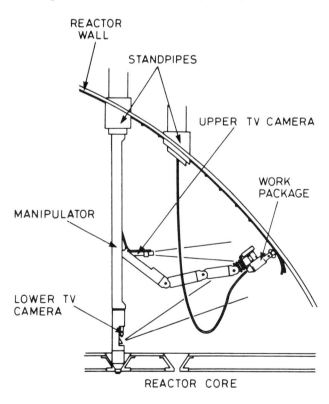

Fig 1 Hydraulic manipulator in position

system is indicated in these cases, though the human operator would still be in overall control. Again the accuracy and stability of the system is limited by the performance of the hydraulic drives.

This paper describes work which is being done in the Remote Operations Section of Nuclear Electric's Technology Division to optimise the performance of existing hydraulic manipulator systems. The overall aims of the work are twofold: firstly, to improve the accuracy and ease of manual joint control; secondly, to provide an accurate and secure control system to which computer guidance packages such as resolved motion and route following may be added. The work is presented as a case study in the use of digital control to improve the performance of existing, relatively unsophisticated equipment to an extent which would otherwise be impossible without extensive hardware upgrading or replacement.

Section 2 describes the hydraulic systems contained in the manipulators. Sections 3 and 4 describe the two different types of power pack in current usage. Section 5 describes a digital controller capable of implementing a variety of algorithms for optimising joint performance. Section 6 describes a digital controller application which is complete and working in the field. Section 7 outlines an application currently under development in the laboratory and section 8 draws some conclusions.

2. MANIPULATOR HYDRAULICS

The actuation system for each manipulator joint is entirely separate. When the manipulators were originally designed, separate pump actuation was adopted as the most economic and easily maintainable option then available, given that at that time very precise performance was not sought. For each joint there is a pump unit connected by a pair of hydraulic lines to a double acting cylinder. The cylinder and piston are annular in shape, so that the whole assembly is hollow, allowing the passage of hydraulic and electrical services to joints lower down the manipulator. Because of the mechanical arrangement, the areas of the up and down pistons area differ. Piston sealing is by means of sliding 'O' ring type seals set in grooves. The piston's linear movement is converted to rotation using a helical spline arrangement. The rotation is used as it is for a 'rotate' type joint, or converted by means of a pair of bevel gears to a 'bend' movement. A typical joint assembly is shown in Figure 2.

The pumps are located together at the top of the manipulator. Each communicates with its joint by means of a pair of 6 mm lines, each of which may be anything up to 15 metres long. These hydraulic lines are partly flexible, consisting of a braided steel sheath lined with PTFE, and partly of rigid copper or stainless steel.

The design of the hydraulic system gives rise to the following main effects, significant from a control point of view:

a) Compliance effects, caused by the flexibility of the hoses and the compressibility of the oil. The effective volume of oil in the system varies with joint position, giving rise to some variation in compliance.

b) Pressure drops in the pipes.

c) Non-linear friction force from the 'O' ring seals. In particular there is a significant 'stiction' force to overcome before the piston begins to move. This combined with the compliance means that there is a small delay after applying a signal before the joint begins to move.

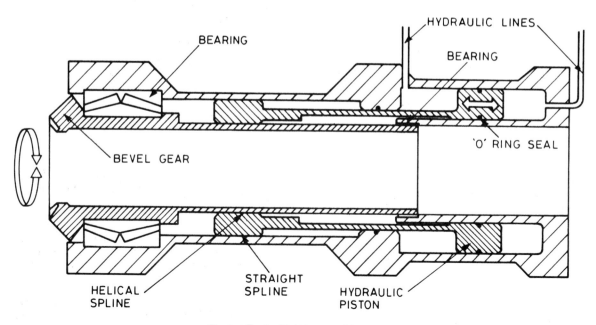

Fig 2 Typical joint assembly

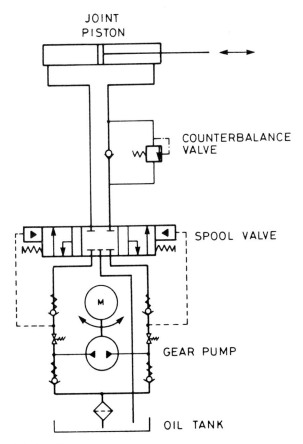

Fig 3 Oildyne drive schematic (simplified)

3. OILDYNE DRIVE

Figure 3 shows schematically the Oildyne pump system. It consists of a gear pump operating a spool valve, which opens to allow flow in one direction or the other when the pump begins to operate. The valve closes to seal the system against leakage when the pump stops. Counterbalance valves are fitted to either one or both lines to prevent cavitation and run-away when driving with load. The gear pump revolves at high speed (up to 10 000 rpm), and is driven by a reversible 110 V universal AC motor. Power is supplied by means of a triac based drive system.

The following characteristics of the Oildyne drive system are significant from a control point of view.

a) The drive system is significantly non-linear, giving rise in particular to a deadband around 0 V input.

b) The gear pump is subject to back leakage. This means that a certain pump velocity is required (in some cases over 1000 rpm) before the joint begins to move at all, further contributing to the deadband.

c) The joint speed is dependent upon load, since there is no velocity control of the pump.

d) Due to the inertia of the rapidly revolving motor, the joint may continue to move for a brief period after the drive power is removed.

Oildyne pumps are used in almost all manipulators in current use within Nuclear Electric.

4. PISTON PUMP DRIVE

Figure 4 shows the piston pump system. This system was developed in order to address some of the shortcomings of the Oildyne unit. At its heart is a radial piston pump with five pistons which are depressed sequentially by a cam mounted on a motor shaft. A system of check valves ensures that the oil is drawn from the tank and directed into the system. A feature of this type of design is that it pumps oil in the same direction, irrespective of the direction of pump rotation. External solenoid operated valves are provided to control flow direction in the system. The piston pump system also includes improved counterbalance valves which sense pressure in the 'outward' line and use it to control the opening of the return line. The pump is driven by a reversible DC servo motor with tacho unit attached, and this is powered by a servo amplifier. The signals for the direction change valve solenoids are derived electronically from the polarity of the drive signal.

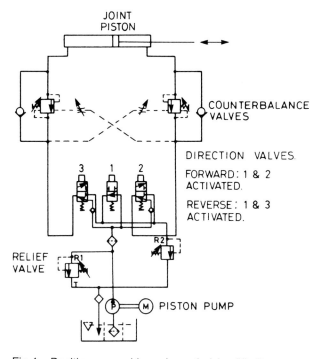

Fig 4 Position pump drive schematic (simplified)

The following characteristics of the piston pump system are significant from a control point of view.

a) The output flow rate is an accurately linear function of input voltage. In particular the pump may be made to pump extremely slowly, and responds to very small input signals.

b) Since the servo amplifier provides closed loop velocity control, the flow rate is entirely independent of load.

c) Because of the closed loop velocity control, there is no inertial run-on after removing the control signal.

Piston pump drives have so far been fitted only to the most recently built manipulator.

5. THE DIGITAL CONTROLLER

Figure 5 shows schematically the control system used for an Oildyne powered manipulator. The joystick signal is fed to a controller board, which in turn feeds an opto-isolator, the purpose of which is to isolate the control signals from the 110 V AC driving the motor. The controller board originally designed for the system was implemented with analogue electronics. It had a dual gain amplifier designed to remove most of the deadband around 0 V, together with circuitry to handle drive end stops and interlocks. Also included was a feedback amplifier which applied PID terms to the signal from the manipulator joint feedback potentiometer, with the aim of providing closed loop position control. Although the analogue boards had extensive use in the field, it had long been realised that some aspects of their operation were limiting manipulator performance. In particular, the analogue electronics were difficult to set up accurately, and this needed to be done uniquely for each joint. Furthermore, the performance of the closed loop position control was far from adequate for providing operator confidence.

In order to address these issues and provide a versatile platform for the control of manipulators and other equipment, a digital controller card has been designed. The design brief was to design a card which was pin-for-pin compatible with the existing board, in order that upgrading could be

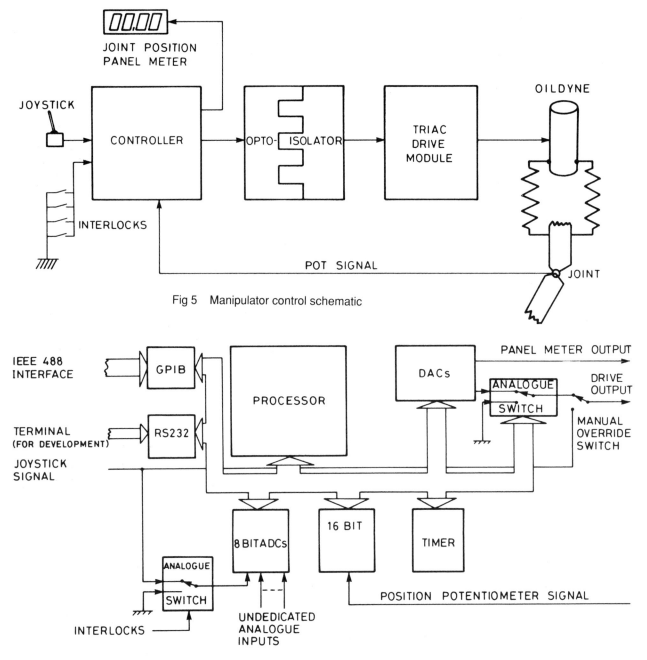

Fig 5 Manipulator control schematic

Fig 6 Digital controller schematic

achieved without further modification to the system. Because of this compatibility requirement, the board was in-house designed. It is however based around a proprietary single board microcontroller. A block diagram of the digital controller is shown in figure 6. The following features are included:

- 16 bit ADC to read the joint potentiometer signal. A high degree of resolution is required, since the joints need extremely accurate positioning in order to obtain acceptable tip accuracy.

- a 16 bit DAC. This output is currently used to drive the existing analogue joint position meters. At some future date, these may be replaced by a fully digital position display system, freeing the output for other uses.

- 8 8 bit ADCs. One of these is used to read the drive joystick.

- 3 8 bit DACs. One of these provides the drive signal to the joint pump motor.

- An IEEE 488 communications interface, allowing digital communications between a number of controllers and a

  host guidance computer or graphics machine.

The CPU is an INTEL 80C196 running at 12 MHz. For programming this, the FORTH language was selected, mainly on the grounds of execution speed and ease of testing.

6. APPLICATION 1: POSITION HOLDING CONTROLLER

Some manipulator tasks require careful positioning of the manipulator tip whilst carrying a heavy load. If the work being carried out is complex and time consuming it may be necessary to maintain position accurately for a period of hours. The original manipulator control system operates in an open loop velocity control mode. Thus it is vulnerable to leakage, caused for example by worn joint piston O rings or a worn Oildyne spool valve, and to movement caused by load changes. Both these effects are quite small, but when the length between joint and manipulator tip is 5 metres, an angular change of 0.1 degree moves the tip by some 8.7 mm, which may well be unacceptable.

A solution would be provided by a simple closed loop position controller operating when the joint is supposed to be stationary and compensating for any load changes or leakage. The impetus to develop the digital controller was in fact provided by the need for this on a specific in-reactor inspection project using an Oildyne powered manipulator: the existing analogue controller board was unable to provide the long-term position stability required.

Figure 7 shows a control block diagram for the position holding scheme developed. Whilst the manipulator is being driven, closed loop control is not required, and the suitably shaped joystick signal provides the drive. The system senses when the joystick is returned to its zero, and, after a short delay, control switches to the position loop. The position signal is obtained from the joint position potentiometer, via the 16 bit ADC. As the position controller switches on, the current position is logged as the datum. The datum remains unchanged until the operator next moves the joystick, and provides the set point to which the position controller drives.

A frequency response based method was adopted for the development of a position loop design. An estimate was made of the amount of loop gain required to achieve the required position stability. Since the joint velocity is proportional to input signal, in the context of position control, this may be considered a type 1 system, and, if linear, should eventually reach the required position. In this application there is no requirement for a quick response. The forward loop gain required was therefore

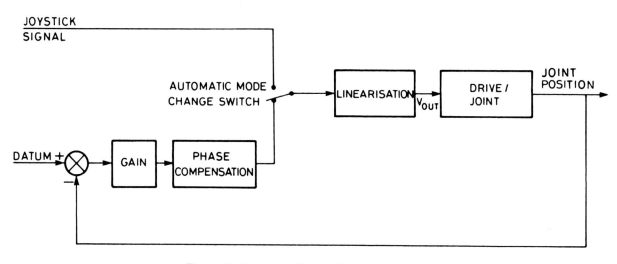

Fig 7 Position controller block diagram

chosen on the basis of the amount of control signal likely to be required to correct an error of a given size, in view of the non-linearity of the drive.

A swept sine frequency response test was performed on a representative manipulator joint mounted in the laboratory. The resulting frequency, gain and phase data were analyzed using the MATLAB matrix manipulation program. A standard method was used to develop a double phase compensator providing adequate gain and phase margin whilst maintaining the zero-frequency gain level at that required.

Ideally frequency response tests should have been carried out on each joint of the manipulator to be controlled. However, limited project time made this an unrealistic option, and the compensation time constants derived for the laboratory test arm were applied to all joints. The view that the test arm would prove a sufficiently representative model of each manipulator joint has been borne out in practice, and it has been found completely adequate in almost all cases to tune the position controllers merely by slightly adjusting the gain settings.

The dual gain amplifier in the original analogue controller board is modelled digitally with a shaping function applied to the drive signal immediately prior to outputting it. The drive deadband is minimised by compensating for it with a region of high gain around 0 volts. This has the effect of both linearising the joystick response, and, by linearising the position loop, improves position hold accuracy and stability.

The program is set to run at a sampling rate of 120 Hz. The complication of an interrupt driven sampler was avoided: the software completes its calculations each cycle well before the next sample is due, then simply polls an on-board timer until the next sample instant is reached. The drive signal is then output and the position and joystick signals read. A number of safety features are built into the program. There are two separate watchdog systems: each has to be accessed by the software at regular intervals, otherwise it will 'time out' causing the system to reset and the drive output to shut down. Shut-downs will also be caused if the angle measured during position control is more than a small distance from the datum, or if the angle measured is significantly outside the end stop limits.

The controller system has recently been commissioned for use on an eight jointed manipulator in the inspection application mentioned to above. Results are very promising, with position stability of typically +/- 0.03 degrees, as compared to a previous typical stability using the position feedback function of the old analogue boards of some +/- 0.5 degrees.

7. APPLICATION 2: HIGH PERFORMANCE TRAJECTORY FOLLOWING

In some applications, maintaining a steady position over an extended period is insufficient. An example is remote welding. In this case, the manipulator tip is required not only to travel to a specific position, but to do so along a specific trajectory. The methods of coordinating the travel of several joints to achieve this are well established, and Nuclear Electric has been carrying out remote welding using an electrically driven manipulator for a number of years (1). Welding with hydraulic manipulators has so far not been possible because of the very limited closed loop position control ability of the available control system. However, with the complementary development of the piston pump and the digital controller, hydraulic manipulator welding is within reach. The project outlined here is aimed at providing a control system capable of allowing a piston pump driven manipulator to carry out remote welding or tasks requiring similar precision.

Again a frequency response based design approach has been adopted. The main difference is that the set point is now moving rather than fixed. Because of the integrating action of the joint and its drive, the position error at any given moment is proportional to the demanded velocity and inversely proportional to the gain. The position loop gain is thus related to the maximum permitted instantaneous position error and the maximum velocity. Again a swept sine frequency response test was carried out using the laboratory test arm, and the response analyzed to develop a suitable phase compensator to provide the required stability.

A complication is that the phase compensator includes a low frequency lag term. While this is not a pure integrator, it makes the system subject to the effect known as 'wind-up'. If a large or rapid position change is required, greater than the ability of the drive to keep up, the cumulative error whilst the joint is 'catching up' can cause severe overshoot. With a digital controller, dealing with this effect is straight forward (2).

The control strategy has been tested with a laboratory model using the same single joint test rig mentioned earlier. Whilst work is at a fairly early stage, results so far are encouraging. The strategy will be implemented using the digital controller. At that point, much of the existing and tested software can be re-used, bringing about savings in development and support costs.

With the versatility of a software based system, the control algorithm possibilities are limited only by the available computing power. The inclusion of the IEEE 488 communications interface allows the fast passing of data both between a joint controller and the guidance host computer and amongst the controllers controlling the various manipulator joints.

This will allow the adoption of more complex schemes:

- feed-forward of manipulator overall configuration to allow individual joint controllers to adjust tuning to compensate for varying gravitational load

- and the 'soak-up', where allowed by configuration, of position errors from less agile joints by those which are more responsive.

In short, the use of software based joint controllers greatly increases the range and sophistication of control strategies which may be implemented, leading to the expectation of a far higher performance level than was previously possible.

8. CONCLUSIONS

The existing analogue joint controller cards in a manipulator control system have been replaced with digital controllers. Without further modification to the control system, a large increase in performance has been made possible:

- the non-linear characteristics of the drive may be more accurately compensated for, giving a much smoother response in open loop velocity control.

- the use of accurate and stable closed loop position control has become possible, where previously performance was inadequate.

- input signals may now be fed to the control system in digital form, cutting down on noise related problems, and suitable for use with a host computer running a manipulator guidance algorithm.

- set-up procedures are greatly simplified by the ability to store tuning settings in the software.

- the digital controller hardware and software provides a solid base with which to attack other, more complex applications. These would include further improvements to joint control, for example closed loop velocity control and self tuning algorithms capable of adjusting to joint load changes, force feedback and sensor integration.

This project has demonstrated the benefits of applying mechatronic principles to the modification of an existing manipulator control system. A considerable increase in performance has been achieved without great expense, and advances not previously feasible may now be contemplated.

In the laboratory clear development routes are now established to increase the performance of the existing manipulator applications, and to apply the system in other contexts. In the field, a manipulator equipped with digital controllers is now working in a Magnox power station reactor pressure vessel, and digital controllers will be fitted to two more items of remote operations equipment within Nuclear Electric during summer 1992.

ACKNOWLEDGEMENT

This work was carried out at Nuclear Electric's laboratories at Berkeley, Littlebrook and Gravesend, and is published with the company's permission.

REFERENCES

(1) THOMPSON, V. R. and JERRAM, K. The Design and Development of Project WARRIOR Equipment. <u>ANS International Topical Meeting: Remote Systems and Robotics in Hostile Environments</u> March 1987, (ANS Inc. Grange Park, Illinois 60525, USA)

(2) ASTROM, K. J. and WITTENMARK, B. <u>Computer Controlled Systems</u>, 2nd edition, 1990, P. 224-226 (Prentice Hall)

Active control of surge in a working gas turbine

M HARPER, BSc, PhD, **D J ALLWRIGHT**, MA, PhD,
Topexpress Limited, UK
J E FFOWCS WILLIAMS, MA, PhD, ScD, FFE, FIMA, FCPS, FASA, FRAS, FloA, FInstP, FAIAA, Hon. Mem., AAAS, FRSA,
University of Cambridge, UK

SYNOPSIS

Surge is a violent instability of the flow through a gas turbine engine which can result in serious damage if unchecked. Surge is therefore avoided at all costs, which limits the available operating range of such engines. We report here on research into means of suppressing the instability by the use of active feedback systems. These detect unsteady pressures in the engine, process them to extract information on nascent instability, and introduce small, rapidly modulated airflows to suppress it. Simple experiments on a working engine have demonstrated significant benefit in available operating range obtained using minimal feedback power.

NOTATION

a	constant
b	constant
k_1	design constant in axisymmetric controller
M	maximum slope of P
n	angular order of flow harmonic
$P_{sa}(V_a)$	compressor characteristic
p, p_s	plenum pressure
p_E	equilibrium value of p
p_T	ambient pressure
R_1	rate of extraction of energy from unsteady flow
R_2	rate at which compressor works on unsteady flow
T	kinetic energy of unsteady flow
t	time
V(t)	mean axial flow velocity
$V_a(\theta, t)$	axial flow velocity
V	equilibrium value of V
$V_t(p)$	exit velocity from plenum
v_n	magnitude of nth angular harmonic of flow velocity

1. INTRODUCTION

Surge has always been a major constraint on the operation of gas turbine engines (Greitzer, 1976). In general terms, what happens is that as load on an engine is increased, the rate of mass flow of air through it decreases. Eventually this leads to a drop in compressor efficiency: the compressor stalls. In axial compressors there can be an abrupt and very large drop in the pressure rise produced by the compressor; the hot, high pressure gas in the combustion chamber can then reverse its flow and rush out through the compressor. The consequent abrupt change of load and temperature is very likely to damage a high-performance compressor.

In centrifugal compressors the process is thought usually to be more gentle. Figure 1 sketches typical performance curves for a centrifugal and an axial compressor. Rather than an abrupt discontinuity, the centrifugal compressor is expected to exhibit a change of sign of the slope of its characteristic.

Greitzer (1976) and Epstein et al (1986) have suggested a simple model of the compressor system and plenum as a Helmholtz resonator, whose eigenvalues' real part is controlled by the slope of the compressor characteristic. At high mass flows, a negative slope discourages oscillation of the axisymmetric flow, while at low mass flows, the positive slope feeds energy into unsteady flows. Consequently these grow exponentially until a limit cycle is reached: axisymmetric surge.

Epstein et al (op. cit.) suggested that active feedback could be used to damp the oscillations, and so achieve stability at lower mass flows. This idea was put to the test on laboratory rigs using turbochargers with centrifugal compressors in Cambridge (Ffowcs Williams and Huang (1989)) and at MIT (Pinsley et al (1990)). Success was reported in both cases. However, Cargill (1990) expressed scepticism as to whether such an idea could work on the multistage, high-speed axial compressors typical of modern high-performance engines, in view of their more violent dynamics, in particular their propensity to pass rapidly into severe stall.

The response must be to look more closely at the process of compressor stall which leads to such behaviour. Moore & Greitzer (1986, I and II) propose a model in which stall is precipitated by circumferential modes of oscillation of the flow around the compressor disc, in much the same way that axisymmetric surge might be precipitated by the Helmholtz mode. Epstein et al (op. cit.) therefore suggested that the same approach - active feedback control - might be applied to prevent compressor stall as well. Obviously the positioning and type of sensors and actuators might be different, but the principle of increased stability achieved by feedback is the same. Pioneering work on the control of stall in laboratory compressor rigs has recently been reported by Paduano et al (1991) and Day (1991). The work of Greitzer (op. cit.; I and II), of Greitzer and Moore (op. cit.; I and II) and latterly of McCaughan (1988, 1988, 1990) suggests that surge and stall are not necessarily independent phenomena. Depending on the geometry of a particular engine and on its operating condition, surge may occur on its own with no non-axisymmetric flows associated with compressor stall. This is known as "deep surge". At the other extreme, stall may occur without producing the axisymmetric oscillations of surge. Alternatively the two may interact, with episodes of rotating stall provoking oscillations in combustion chamber pressure in a phenomenon referred to as "classic surge".

In this paper we present the results of a series of experiments on a working engine, and some of the theoretical and numerical modelling work which these results have stimulated. Initially, the work was expected simply to reproduce the results of Ffowcs Williams and Huang (op. cit.) on a working engine; in practice things have not turned out so simply. In section 2 we present some results obtained from our experiments so far; in section 3 we present our current ideas on the most appropriate feedback control strategy for surge and stall.

2. EXPERIMENT

Our experiments have been carried out on a Rover IS/60 45kW auxiliary power unit. This has a single-stage centrifugal compressor with a pressure ratio of 3:1; a single reverse-flow combustor, and a single stage axial turbine. A mechanical fuel governor maintains a virtually constant compressor delivery pressure versus load at high loads. A section of the engine is shown in Figure 2. A set of four pressure sensors are located in the casing wall around the impeller eye. A second set is located at the inlets of four out of nine diffuser channels; and a third set of four are situated around the combustion chamber.

2.1 Control of Axisymmetric Surge

In an experiment reported by Ffowcs Williams and Graham (1990), an attempt was made to repeat the laboratory experiment of Ffowcs Williams and Huang (op. cit.) by damping the Helmholtz mode of oscillation. Pressure signals from the combustion chamber were fed to a high-Q, analogue electronic filter tuned to what was believed to be the Helmholtz resonance frequency. The output from this was fed to a high-speed proportional air valve which fed compressed air back into the combustion chamber.

This arrangement was not found to be able to extend the range of stable operation. They did however succeed in getting their control system to destabilise a form of 'metastable' surge induced by large transient overfuelling of the engine at low load. The control system delivered a carefully-timed pressure impulse to the combustion chamber, knocking the engine out of surge and back into normal, low load operation once the overfuelling was removed.

Our first development was to construct a feedback loop using a broadband digital filter. This was programmed to provide proportional feedback over a broad frequency range of 0.1 - 100Hz: the surge limit cycle fundamental frequency was about 12Hz. It was implemented using a proprietary digital signal processing board mounted in a PC-compatible computer. The board was an ADSP2100 system board supplied by Loughborough Sound Images Ltd. It provided a 200-tap finite impulse response filter running at a sampling rate of approximately 20kHz. Otherwise the control system was as described by Ffowcs Williams and Graham (op. cit.).

The effect of applying feedback was intriguing. At loads at which the engine would normally surge continuously, it would continue stable operation, but punctuated by occasional "blowdowns" of plenum pressure. These are effectively half of a surge cycle, and are followed by a rapidly-dying oscillation at the surge frequency, as shown in Figure 3.

These "blowdowns" are immediately preceded by strong oscillations in static pressure in the compressor's diffuser channels. These occur at 175 - 190Hz (about one-quarter of shaft rate) and can exceed 4PSI peak-to-peak. The oscillations are coherent between diffuser channels; the relative phase between channels indicates a bimodal disturbance rotating at half of 190Hz. This is typical of rotating stall dominated by the second angular harmonic.

These oscillations also precede normal, uncontrolled surge, as can be seen in the lower trace of figure 5, which shows a diffuser trace around the onset of surge at about 0.14 sec. We therefore interpret our observations as follows. Our centrifugal compressor is behaving rather as axial compressors are expected to: it is stalling abruptly, producing blowdown of the combustion chamber. Our simple, single-channel control system could not prevent this; however, once the compressor recovered, the controller acted to damp the surge oscillation very quickly, as can be seen from Figure 3. After a random time-interval the stalling process is then re-triggered by the very high levels of pressure-noise present in the engine (typically 1PSI in the combustion chamber). This interpretation is shown schematically in figure 1: noise causes the instantaneous operating point to move around within the vicinity of the time-average operating point. As the latter approaches the stall point, but before it reaches it, noise will trigger stall and hence surge. This interpretation also explains the experiment observation that the compressor stalls before the slope of the characteristic changes sign.

When load on the engine was further increased, it was observed that blowdowns rapidly became more frequent: clearly this method of control was not of itself going to increase the available operating range very much.

2.2 Control of Unsteady Flow in the Compressor

In the next stage of our work we therefore turned our attention to compressor stall. The oscillations described above are always seen in the diffuser at high engine load: the solid line in Figure 6 shows a typical pressure spectrum (arbitrary amplitude units). The resonance becomes stronger, narrower and moves up slowly in frequency as engine load increases towards the surge point: this behaviour is just what we expect as a mode tends towards instability. We applied feedback in order to increase the modal damping and so postpone the onset of stall.

The control system is shown in outline in Figure 4. A number of pressure sensors are positioned flush in the walls of diffuser channels around the compressor. Their signals are conditioned and routed to a many-input, single-output controller. This provides a control signal to a high-speed actuator, which modulates a flow of compressed air. The actuator is essentially a lightweight sleeve valve driven by a loudspeaker coil and has a useful bandwidth well in excess of the required figure of 200Hz. The modulated flow is fed to a point at the outer edge of the impeller eye, where it is injected parallel to the impeller axis. In typical control experiments, with the engine at the uncontrolled surge line the RMS level of modulation would be very roughly 0.14% of the engine mass flow.

The controller is essentially a multi-channel digital FIR filter implemented using the same hardware as before. The controller is capable of providing a four-in one-out filter matrix with 64 coefficients per channel at a sampling rate of 20kHz.

The practical implementation of control required signals in the feedback loop to be band-limited to reduce noise. They also had to be phase-shifted: individual sensor signals had to be treated to produce an efficient mode-observer; the output signal had to be treated to compensate for the effect on phase of delays in the actuation system. The chosen method was to incorporate further delays, which avoided distortion of signal power spectra and allowed easy and rapid adaptation of the controller response. However, control systems incorporating delays are known to be limited in the stability margin which they can produce (Marshall 1979).

Subsequent modelling has suggested this was probably a serious limitation in this case.

Response before Surge

The effect of feedback on the spectrum of a diffuser pressure signal is shown in Figure 6. The spectrum was recorded during normal operation, with engine load too low to precipitate surge. The developing instability in the absence of control (solid line) produces the broad peak centred at 190Hz. On applying feedback control (dashed line) the peak is reduced, is broader and has moved towards lower frequency: the resonance has become more heavily damped.

Control Trials

Trials consisted of testing whether control had moved the engine's surge point, using three different techniques. Our most commonly-used measure of performance was the engine power delivery to the dynamometer just before surge. The engine governor was mechanical, and maintained a constant pressure ratio over the range of (stable) operating conditions used.

Test 1: The engine's surge point would be established without control by slowly increasing load via a water-brake dynamometer. The controller would then be switched on and the surge point remeasured using the same procedure. In practice the surge point tended to drift: it could differ by 1% in engine load between 2 measurements in rapid succession, and by 3-4% over the course of a day's testing. Our tests were therefore done in as rapid succession as possible, but occasionally there would be a delay of up to 20 minutes. In view of this we suggest there is an uncertainty of ±2% in any measurement of power at surge (measured as being proportional to dynamometer load times shaft speed). Table 1 shows results from a succession of control filter settings. In filter set one, a first-order filter function was used to shape the frequency response, while in set 2 a second-order filter function was used. Phase was optimised by adding or subtracting delays: letter subscripts denote different delays. The precise phase shifts required for optimum control had to be approached by trial and error. An initial set of values would be calculated using the transfer functions from actuator drive to pressure sensor signals measured with the engine at a stable operating point. The transfer functions were known to change with the operating point, and hence these values would not be optimal when the engine was taken up to and beyond the normal surge line. Changes of up to 40° were required to obtain the best performance.

TABLE 1

Surge Margin Achieved by Control: Test 1

Filter Set Number	Power Increase %
1a	4
1b	6
1c	5
1d	4, 1.5
1e	6
1f	8.5, 9, 5.5
1g	11.5, 7
2a	6
2b	3.5

The variability of power at surge is clear from the results of repeated trials with filters 1d, 1f and 1g in Table 1. Corrected mass flow through the compressor at surge appeared to vary much less than delivered power and could be deduced from static pressure measured in the throat of a bellmouth on the air intake. This was occasionally recorded on tape. For trials with filter set 1g, corrected mass flow was 2.6% lower at surge with control on than with control off.

Test 2: The engine surge point without control would be determined; control would be switched on and engine load increased to about 5% above this point. Control would then be switched off. This was tried 4 times. On 3 occasions the engine surged immediately control was removed; on one, it surged while control was still on.

Test 3: The engine would be surged without control. Control would then be turned on to see if surge was then suppressed. This was tried twice without success. The reason became obvious on analysing recordings of the actuator's displacement: the control system responded to some extent to the pressure variations produced by the very large changes in axisymmetric flow during surge. These saturated the control system and prevented it responding properly.

3. A COMBINED STALL/SURGE CONTROL SYSTEM

It is clear from these experiments that control of the total flow alone (Section 2.1) is not sufficient. The local unsteady flow in the compressor must also be controlled. This we expect to be true, not just for the test engine, but for any engine in which stall and surge are both possible.

In what follows, we present an outline specification for a combined control system. The theory of instability is far better developed for axial than for centrifugal compressors, and we take this as our starting point. Since the qualitative behaviour of our centrifugal compressor is so strikingly similar to predictions for axial machines, we believe that the results are of interest for both types.

Our outline specification is based on the use of two distinct control systems, corresponding to the two distinct instabilities. In general terms therefore we shall need

(A) A controller on the compressor alone that measures and controls any non-axisymmetric disturbances to the flow.

(B) A controller on the compressor-plenum system that measures and controls any disturbances to total mass flow and plenum pressure.

We shall assume that the compressor is satisfactorily modelled by Moore's theory of finite amplitude disturbances in a multistage axial compressor (Moore, 1984). This theory has been further developed in the joint work of Moore and Greitzer (1986, I and II), and in the work of McCaughan (1988, 1988 and 1990). It is a nonlinear theory that appears to account successfully for many of the observed features of compressor behaviour, as described by Greitzer (1976, II). In particular, the model predicts the existence (depending on parameter values) of steady non-axisymmetric flow, unsteady axisymmetric limit cycles, and unsteady non-axisymmetric limit cycles, which are the observed features of rotating stall, deep surge, and classic surge respectively. So we shall analyse theoretically the effects of applying suitable control systems to this model.

Controller A

Let $V_a(\theta, t)$ be the axial velocity through the compressor at position q round the circumference at time t. We write its Fourier series as

$$V_a(\theta, t) = V(t) + \sum_{-\infty}^{\infty}{}' v_n(t) e^{in\theta} \qquad (1)$$

where the ' on the summation indicates that the term $n = 0$ is omitted. Then we include our controller term $k(V_a - V)$

into Moore's equation for the pressure rise from ambient (p_T) to plenum (p_s):

$$p_s - p_T = P_{sa}(V_a) - a_0 \frac{dV}{dt} - \sum_{-\infty}^{\infty'} a_n \frac{dv_n}{dt} e^{in\theta} - b \frac{\partial V_a}{\partial \theta} - k(V_a - V) \tag{2}$$

Here P_{sa} is the assumed steady axisymmetric characteristic, and a_0, a_n and b are constants depending on the physical parameters of the compressor and ducts. When we multiply this equation by $V_a - V$ and integrate over q from 0 to 2π, the axisymmetric terms produce 0 because V is the mean flow. Also

$$\int_0^{2\pi} \frac{\partial V_a}{\partial \theta}(V_a - V)\,d\theta = \left[\frac{1}{2}(V_a - V)^2\right]_{\theta=0}^{\theta=2\pi} = 0. \tag{3}$$

So we are left with

$$0 = \int_0^{2\pi} P_{sa}(V_a)(V_a - V)\,d\theta - 2\pi \sum_{-\infty}^{\infty'} a_n \frac{dv_n}{dt}\bar{v}_n - k\int_0^{2\pi}(V_a - V)^2\,d\theta \tag{4}$$

We write this in the form

$$\frac{dT}{dt} = R_2 - R_1 \tag{5}$$

where $\quad T = \sum_{-\infty}^{\infty'} \pi a_n |v_n|^2 \tag{6}$

measures the kinetic energy of the non-axisymmetric flow,

$$R_2 = \int_0^{2\pi} P_{sa}(V_a)(V_a - V)\,d\theta \tag{7}$$

measures the rate at which the compressor is working on the non-axisymmetric flow, and

$$R_1 = k \int_0^{2\pi} (V_a - V)^2\,d\theta \tag{8}$$

measures the rate at which controller A is reducing the kinetic energy of the non-axisymmetric flow.

Now since $V_a - V$ integrates over q to zero, and V is independent of q,

$$R_2 = \int_0^{2\pi}(P_{sa}(V_a) - P_{sa}(V))(V_a - V)\,d\theta = \int_0^{2\pi} \frac{P_{sa}(V_a) - P_{sa}(V)}{V_a - V}(V_a - V)^2\,d\theta. \tag{9}$$

If M is the maximum positive slope of P_{sa} then the difference quotient in this last form of R_2 is always at most M, so

$$R_2 \leq M \int_0^{2\pi}(V_a - V)^2\,d\theta \tag{10}$$

So if the controller gain k exceeds M then we have $R_2 \leq R_1$ and so the controller is always reducing the energy in the non-axisymmetric part of the flow faster than the compressor is increasing it (except, of course, when $V_a \int V$ i.e. when the flow is axisymmetric). This argument is perfectly valid with a time-varying V and so does cover exactly the requirement for controller A : the departure of the flow from axisymmetry is steadily reduced to zero, whatever the mean flow is doing.

Controller B

Following Moore's theory again, the uncontrolled equations for the plenum pressure p and flow velocity V will be

$$\frac{dV}{dt} = a(P_{sa}(V) - p) \tag{11}$$

$$\frac{dp}{dt} = b(V - V_t(p)) \tag{12}$$

for some positive constants a and b, with $V_t(p)$ being proportional to the steady exit velocity from the plenum when the pressure is p. At the required steady equilibrium point, the values p_E and V_E of p and V satisfy

$$p_E = P_{sa}(V_E) \tag{13}$$

$$V_E = V_t(p_E) \tag{14}$$

The matrix of the linearized equations about equilibrium is

$$\begin{pmatrix} aP'_{sa}(V_E) & -a \\ b & -bV'_t(p_E) \end{pmatrix} \tag{15}$$

A controller using direct proportional feedback on plenum pressure would replace the pressure equation by

$$\frac{dp}{dt} = b(V - V_t(p)) + k_1(p - p_E) \tag{16}$$

The investigations carried out so far along these lines indicate that a control system designed in this way is capable of not only stabilizing the equilibrium point to small disturbances, but also eliminating the possibility of unsteady axisymmetric behaviour of finite amplitude. Thus it appears that there is no fundamental reason why the requirements for controller B should not be met in this way.

4. CONCLUSIONS

Separate experiments on the control of axisymmetric surge and of rotating stall in a working engine show that both are able to suppress surge, at least over a limited range of operating conditions. Since surge behaviour in the test engine is produced by an interaction of these two instabilities, both forms of control will be required to obtain worthwhile gains in performance.

An analysis based on axial compressor theory shows that a pair of linear controllers applied to surge and stall have the potential to guarantee stability, even against transients in operating conditions which would otherwise lead to surge.

5. ACKNOWLEDGEMENTS

This work has been funded under a series of grants from the US Office of Naval Research.

Significant contributions to the work have been made by N Piercy[1], W Hodson and D Williams of Topexpress Ltd. We also express our appreciation of the assistance of H Freeman, R Oliver and M Dredge of Noel Penny Turbines Ltd, at whose Coventry site the engine was tested.

6. REFERENCES

[1] CARGILL, A M, and FREEMAN, C, (1990).
High Speed Compressor Surge with application to Active Control. ASME Gas Turbine and Aeroengine Congress and Exposition, June 1990. Paper Number 90-GT-354.

[2] DAY, I J, (1991)
Active Suppression of Rotating Stall and Surge in Axial Compressors. ASME Gas Turbine and Aeroengine Congress and Exposition, June 1991. Paper Number 91-GT-87.

[3] EPSTEIN, A H, FFOWCS WILLIAMS, J E and GREITZER, E M.
Active Suppression of Compressor Instabilities. AIAA 10th Aeroacoustics Conference. Paper number AIAA-86-1994.

[4] FFOWCS WILLIAMS, J E and GRAHAM, W R.
An Engine Demonstration of Active Surge Control. ASME-IGTI Congress and Exposition, June 11-14, 1990, Brussels. Paper number 90-GT-224.

[5] FFOWCS WILLIAMS, J E and HUANG, E Y.
Active Stabilisation of Compressor Surge. J Fluid Mech 204, 245-262 (1989).

[6] GREITZER, E M.
Surge and Rotating Stall in Axial Flow Compressors: Part I: Theoretical Compression System Model. Trans. ASME Jnl Engineering for Power, 98, 190-198, April 1976.

[7] GREITZER, E M.
Surge and Rotating Stall in Axial Flow Compressors: Part II: Experimental Results and Comparison with Theory. Trans. ASME Jnl Engineering for Power, 98, 199-217, April 1976.

[8] MARSHALL, J E.
Control of time-delay systems. IEE Control Engineering Series, Volume 10. 1979.

[9] McCAUGHAN, F E.
Application of Bifurcation Theory to Axial Flow Compressor Instability. ASME Gas Turbine and Aeroengine Congress and Exposition, June 1988. Paper number 88-GT-231.

[10] McCAUGHAN, F E.
Numerical Results for Axial Flow Compressor Instability. ASME Gas Turbine and Aeroengine Congress and Exposition, June 1988. Paper number 88-GT-252.

[11] McCAUGHAN, F E.
Bifurcation Analysis of Axial Flow Compressor Stability. SIAM Journal of Applied Mathematics, 50, 1232-1253, October 1990.

[12] MOORE, F K.
A Theory of Rotating Stall of Multistage Axial Compressors: Part I- Small Disturbances, Part II - Finite Disturbances, Part III - Limit Cycles. Transactions of ASME Journal of Engineering for Gas Turbines and Power, 106, 313-336, April 1984.

[13] MOORE, F K, and GREITZER, E M.
A Theory of Post-Stall Transients in Axial Compression Systems: Part I - Development of Equations. Transactions of the ASME Journal of Engineering for Gas Turbines and Power, 108, 68-76, January 1986.

[14] GREITZER, E M, and MOORE, F K.
A Theory of Post-Stall Transients in Axial Compression Systems: Part II - Application. Transactions of the ASME Journal of Engineering for Gas Turbines and Power, 108, 231-239, April 1986.

[15] PINSLEY, J E, GUENNETTE, G R, EPSTEIN, A H and GREITZER, E M.
Active Stabilisation of Centrifugal Compressor Surge. ASME-IGTI Conference and Exposition, June 11-14, 1990, Brussels. Paper number 90-GT-123.

[16] WONHAM, W M.
Linear Multivariable Control: a Geometric Approach (second edition). New York, Springer-Verlag, 1979.

---

[1] Now at Scientific Generics Ltd, Cambridge

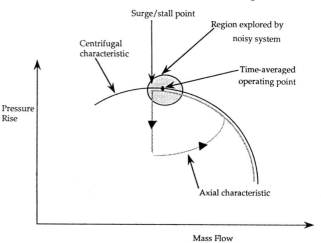

Fig 1 Compressor characteristics expected to be typical for centrifugal and axial compressors

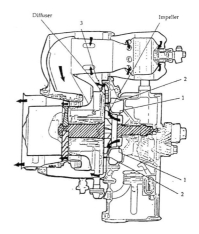

Fig 2 Engine section showing positions of pressure sensors
1 – Four around impeller eye
2 – Four in diffuser passages
3 – Four at compressor delivery

All at 90 degrees intervals around axis

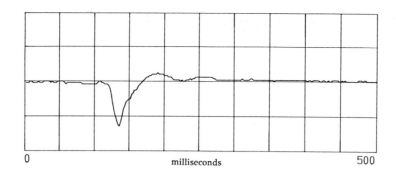

Fig 3   Time history of a 'blowdown' during surge control trial

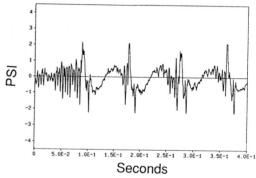

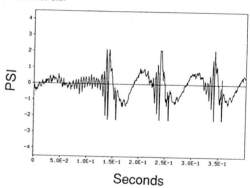

Fig 4   Stall control system

Fig 5   Effect of control on surge mechanism

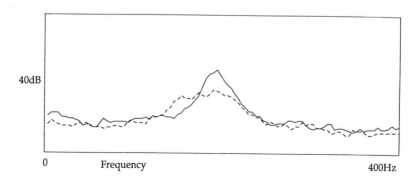

Solid: Control off

Dashed: Control on

Fig 6   Effect of control on diffuser spectrum

Intelligent control of cone crushers

R A BEARMAN, PhD, **R M PARKIN**, BSc, PhD, CEng
Leicester Polytechnic, UK

SYNOPSIS Cone crushers are used in the production of aggregate for use in roadmaking and other industries. The UK aggregate industry produces around 300 MTonnes/year involving an energy consumption of some 6.25 TWh/year from the National Grid.

Traditional control systems for rock crushers rely on observation by personnel and manual adjustments effected mechanically or via a control panel and electrical means.

A mechatronic approach to the design of cone crushers is reported. Control of operating parameters being achieved by electrohydraulic methods facilitates fully automatic processor control. A Knowledge Based System is used to define the optimal crushing regime whose parameters are then maintained via distributed control and condition monitoring. The control system is supported by the feedback of various parameters derived from the use of some innovative sensors. The intelligent monitoring system is used to avoid operational damage to the crusher from foreign objects such as "tramp iron" whilst keeping the product at a uniform specified size. The crusher builds up its own operational data base permitting predictive maintenance and providing useful historical archives to aid in the diagnosis of causes of catastrophic failures. The knowledge base thus acquired is to be used as an input to future design processes.

NOTATION

Y   fraction of cumulative undersize,
n   slope of the Rosin-Rammler-Bennet (RRB) graph at 63.21% cumulative undersize,
X   size fraction (mm)
X_o size fraction at 63.21% cumulative undersize.

1 INTRODUCTION

1.1 Cone Crushers

The cone crusher as seen in Fig 1 is a Pegson Autocone crusher. Cone crushers are manufactured by all the leading mining and quarry plant manufacturers to differing levels of sophistication. The action of a cone crusher relies upon rock entering the crushing chamber and being gripped by the crushing members (concave and mantle) to produce breakage. The eccentric shaft mounted about the main stationary shaft imparts a rocking motion to the mantle which also gyrates at approximately six revolutions per minute.

In terms of basic design there are three types of cone crusher:
i. discharge setting adjusted via screw thread with spring loaded overload protection eg, Pegson Gyracone, Nordberg Gyradisc, Allis Symonds, Kue-Ken CT range
ii. discharge setting adjusted via screw thread with hydraulic overload protection eg, Nordberg Omnicone, Telsmith
iii. discharge setting adjusted via hydraulic means with hydraulic overload protection eg, Pegson Autocone, Automax and Autosand, Allis Hydrocone.

Of the types listed above only (iii) with the fully hydraulic system is capable of utilizing full closed loop control. All the cone crusher ranges offered by manufacturers have alternative concave and mantle designs that give various feed openings and differing degrees of product fineness. Cone crusher sizes are quoted as the diameter of the cone head, these vary from 0.45 metre to 3.05 metre with a concomitant variation in drive motors from 40kW to 750kW.

1.2 Cone Crushers in Mining and Mineral Processing

In quarrying and mining the main aim is to reduce the as-blasted rock to a size suitable for either sale or further metallurgical treatment. Size reduction thus accounts for a significant percentage of total processing costs.

From the mining viewpoint it is the separation and liberation of the valuable minerals from the gangue minerals that is the prime objective. To achieve this objective generally requires a size reduction from 1-2 metre diameter to 0.1-5mm diameter, with final size governed by subsequent treatment. The reduction starts with the primary crusher(s) where the as-blasted rock is crushed in either jaw, gyratory or impact type crushers. The secondary, tertiary and quarternary stages typically employ cone type crushers designed to produce increasingly finer products. The particle size at which crushing ceases and grinding begins is a function of the final recovery process and the nature of the mineral to be extracted. The grinding stage utilizes autogenous, semi-autogenous, ball and rod mills. Grinding is highly energy intensive due to the processes being very inefficient in the transfer of energy. Fig 2 shows a typical mineral processing circuit with primary, secondary and tertiary crushing stages and subsequent grinding and flotation. The mining industry worldwide utilizes many thousands of cone crushers in various mineral processing applications. The importance of cone crushing in mining is purely as a mineral preparation stage. In terms of energy Fig 3 shows the percentages of electrical energy used in the various stages of the treatment of a copper-molybdenum ore. As can be seen from Fig 3 the percentage of energy consumption attributable to crushing is relatively low. Of this percentage probably only 50% is due to cone crushing, thus cone crushing represents only 6% of the energy costs related to mining and mineral extraction. If the final metal extraction and refining costs are also considered then cone crushing energy consumption becomes negligible.

Although efficient control of cone crushers would be an advantage to the mineral processing industry, the obvious inference is that research directed at grinding, extraction and refining is far more expedient.

1.3 Cone Crushers in Quarrying.

The end result desired in quarrying is to produce material for highway, concrete and asphalt production. The requirements of these industries is either single sized aggregate or graded aggregate. The gradation, composition and shapes of these materials is laid by various standards authorities.

A typical quarry flowsheet is shown in Fig 4. Unlike mineral processing, cone crushers are responsible for producing the majority of the final saleable product. In order to produce the final product cone crushers consume on average between 0.8 and 2.0 kwh/tonne per individual machine. Bearman (1) examines energy consumption in the U.K quarrying industry and, based on figures supplied by the British Aggregate Construction Materials Industry (BACMI), estimates that 30% of all energy consumption in quarrying is consumed by crushing. Of that 30% up to 80% is consumed after the primary stage. A vast majority of energy consumed after the primary stage is utilized in cone crushers. As a consequence 20-25% of all energy costs in the U.K quarrying industry are consumed in cone crushing, this equates to approximately 0.75% of the total energy generated in the U.K every year. Fig 5(a) shows the breakdown of energy consumption in the quarrying industry. Figs 5(b) and (c) illustrate that certain sectors of the quarrying industry are even more energy intensive.

Pegson Ltd. are the leading U.K based manufacturer of jaw and cone crushers. In the field of cone crushers they market the established Autocone range and in addition the highly innovative Automax and Autosand attrition based cone crushers. In conjunction with Leicester Polytechnic and Transfer Technology Group plc they are carrying out a research project aimed at improving the efficiency of cone crushers utilising a mechatronic approach. The project has attracted funding from the Department of Trade and industry under their LINK initiative.

2 AIMS OF THE CONTROL SYSTEM AND INITIAL MODEL DEVELOPMENT

The principle aim of the project is to improve the efficiency of cone crusher operation through the application of a mechatronically integrated control system.

It is essential that cone crushers produce a consistent high quality product with optimal energy usage. If cone crushers are not operating efficiently the following effects are noted:

i. product does not reach specification and is either used for inferior (lower value products) or is blended incurring additional handling costs
ii. inefficient crushing can overload classifying screens reducing their efficiency
iii. low efficiency crushing increases the amount of oversize material thus increasing the circulating load within the plant thus increasing handling costs and energy consumption
iv. poor feed regulation to the crushers increases wear costs drastically and increased wear reduces the efficiency of the crushing process and produces particles of poor shape
v. too much fine material can be produced which may represent non-saleable material
vi. poorly shaped unsaleable material is produced if the crusher is not correctly fed
vii. temporary machine overloads occur which can become cyclic, thus causing a decrease in production rates
viii. inefficient operation may lead to an increase in mechanical failures and lost production due to repairs

Operationally the aim of the system is to maintain the optimum discharge setting to obtain the required product. This must be achieved with due regard to the mechanical limits of the crusher and the state of wear of the crushing members. A major factor in the proposed control system is the ability to detect faults. Monitoring the previously outlined parameters the intention is to predict failures.

The basis of the proposed control system is the relationship developed by Bearman (2), who has proven that a relationship exists between the strength properties of rock, the discharge setting of the cone crusher and the performance of the crusher. In his work on rock mechanics, an extensive suite of strength parameters were correlated with energy consumption, product size and gradation. The only test to show significant correlations with energy consumption and product size grading was the fracture toughness test. The fracture toughness test has been used for the testing of metals and composite materials for many years. During the 1980's many workers in the field of rock mechanics attempted to adapt the test to rock (3),(4). In 1987 the International Society for Rock Mechanics issued recommendations (5) regarding the mode of fracture toughness test suitable for application to rock. Bearman used the chevron bend test method as illustrated in Fig 6 This test was applied to a wide variety of rock types used in the production of aggregates. From the correlations it proved possible to construct a three-dimensional equation relating energy consumption, fracture toughness and discharge setting.

The product grading from cone crushers is accurately estimated by the Rosin-Rammler-Bennett (RRB) equation as seen below:
$$\ln(\ln(1/(1-Y))) = n.\ln X - n.\ln X_o$$

The Rosin-Rammler-Bennet (RRB) method requires time consuming sieve analysis and plotting on special graph paper. Bearman thus attempted to predict the descriptors X_o and n. He found that X_o and n can be predicted from the fracture toughness of the rock. If these descriptors are then substituted back into the equation the size distribution can be predicted. The outlined procedure forms the basis of the control algorithm.

3 CONDITION MONITORING AND SOFTWARE REQUIREMENTS

In order to supply the data required by the control system the following parameters will be monitored:
i. discharge setting
ii. power consumption
iii. hydraulic pressure
iv. oil temperature
v. vibration levels and frequency
vi. machine throughput
vii. level of feed
viii. product size and grading
ix. product shape
x. degree of wear on the crushing members
xi. oil flow
xii. rotation sensor underneath mantle

For an explanation of some of the above terms see Fig 1.

The parameters listed above are to be monitored so that the cone crusher continually operates at its peak performance level. A monitoring and control software package is required to handle the data and act as a constantly available, experienced crusher operator. The requirements of any software package are outlined below:
i. accepts a multiplicity of inputs (12 plus)
ii. good database facilities and/or seamless interfaces with external databases
iii. trending capabilities for current and historical data
iv. real-time and consultative capabilities
v. be capable of utilizing empirical data as part of the control
vi. good user interface
vii. flexible programming environment
viii. explain function to stop operator alienation
ix. high quality graphics capability

There is a high degree of empirically derived knowledge attached to crusher operation. The selected software option would need to be capable of storing empirical knowledge and utilizing it during the decision making process. Knowledge Based systems provide many of the functions desirable for this type of control system.

Many early applications of Knowledge Based systems were off-line and stand-alone in nature. In recent years there has been several successful applications of Knowledge Based systems to real-time applications. It is critical to the success of the project that a suitable Knowledge Based system tool is selected, thus several systems have been investigated.

A major developer of Knowledge Based systems for both on-line and off-line projects is Comdale Technologies Inc.. The company markets Comdale /C and Comdale /X shells. Harris and Kosick (6) reported an early application of the real-time Comdale /C system in 1988. The shell had been applied at the Polaris mine to act as a real-time process controller for the flotation circuit. Meech (7) examined the use of Comdale /X in a teaching role and also as a plant simulator and for the assessment of Rock Mass Rating. Comdale /X is the stand-alone version and has been used by Meech in the training of students. Harris (8) gives a list of applications where Comdale has been applied. The list of applications is reproduced below:
- Monitoring and control of a grinding circuit at Cement Lafarge, New Braunfels, Texas,
- Monitoring and control of a kiln at St. Lawrence Cement, Joillette, Quebec,
- Controlling the grinding circuit at Wabush Mines, Labrador,
- Controlling column flotation cells at Minnovex, Technologies Inc., Toronto, Ontario,
- Monitoring and control of a copper flotation circuit at Mount Isa Mines, Queensland, Australia,
- Troubleshooting, diagnosis and metallurgical control of a copper smelter, Mount Isa Mines, Queensland, Australia,
- Troubleshooting and diagnostics of a Katapac operation, Kodak, Toronto,
- Operator flotation assistant, Inco, Manitoba,
- Designing production and service hoisting systems for Dynatec Engineering, Toronto, Canada,
- Monitoring and control of a grinding and flotation circuit at Gibraltar Mines, British Columbia,
- Boiler plant load management at Wabush Mines, Labrador,
- Assisting in the operation of a lead blast furnace, Broken Hill Associated Smelters, South Australia,
- Load haul dump maintenance assistant, Inco, Sudbury, Ontario, Canada,
- Monitoring of a flotation circuit at Hyland Valley Copper, British Columbia,
- Grinding, flotation and dewatering monitoring and control, Minnova Inc., Ontario, Canada.

Documented examples of the use of the Comdale systems include a copper flotation Knowledge Based system at MIM Ltd. (9) and an Knowledge Based system for control of an autogenous mill circuit (10). Harris et al have also produced a paper justifying the use of Knowledge Based systems in the process industries (11).

The SUPERINTENDENT Knowledge Based system shell from Heuristics Inc., was tested at Brenda Mines (12). The SUPERINTENDENT system is a real-time system which is run concurrently with the ONSPEC software supplied by Heuristics Inc.. The ONSPEC software contains modules to handle data input/output, database manipulation, alarm signals and graphics screens. One of the modules within ONSPEC is the SUPERTRENDS package. SUPERTRENDS allows graphical display of current and past values on an adjustable time scale. The knowledge base developed is termed GRINDX and in order to validate the GRINDX knowledge base its performance was compared to the CP250 supervisory control program.

A major competitor to Comdale is the G2 real-time Knowledge Based system supplied by Gensym Corp.. At Brunswick Mining and Smelting a G2 system was added to an existing Distributed Control System (DCS) (13). G2 uses generic rules to control a grinding circuit. The generic rules vastly reduce the time and effort required to develop process control systems. The G2 system used has only 50 rules to balance the material flow in three grinding circuits and the flotation circuits. The integration of Knowledge Based systems and DCS is a powerful option.

Within the cement industry a notable success story is that of the LINKman Knowledge Based system (14). Blue Circle plc are a major producer of cement products. The project ran from 1984-1986 and the aim was to produce a control system for cement kilns. There are twelve process parameters that are routinely monitored for control purposes. Blue Circle and Sira Ltd. developed the process control software which incorporated knowledge from control engineers and rules-of-thumb. The realisation that they had developed an Knowledge Based system arose only after implementation. The system contains 200 primary and secondary rules. The input signals to the LINKman system are exhaust gas levels, exhaust gas temperature, kiln drive power and product chemistry. In response to these inputs the control actions include fuel flow rate, fuel/air ratio, input feed rate, kiln rotation speed, cooling system and exhaust draught.

The LINKman system is in use at six of the Blue Circle plants where the Knowledge Based system provides stable plant operation at the lowest temperatures possible. The benefits claimed include fuel reduction, reduced wear on refractory linings, increased output and energy savings in the region of 2.5-4 million per year.

In the United Kingdom the Knowledge Based system shell which has been most successfully used for applications is Crystal from Intelligent Environments Ltd. This is a PC based, backward chaining, rulebased shell with extensive facilities for screen presentations. The shell is implemented in a very efficient manner, and as a result runs very quickly making it suitable for large applications and on-line or real-time applications. It has proven to be very easy to use and extremely flexible, which explains to a great extent why it has become the most widely used Knowledge Based system product in the UK.

In fact the product is proving itself as a very powerful fourth generation programming environment for the development of a wide range of software applications, in addition to just Knowledge Based systems. It is simple to interface Crystal to a wide variety of spreadsheets, databases and C programs making interfacing for on-line applications very easy.

In industry there are a growing number of companies specialised in developing applications involving Knowledge Based system software. In the United Kingdom, the most successful of these is Intelligent Applications Limited (IA). IA specialise in developing on-line diagnostic systems in the manufacturing and process industries. Some of the projects successfully undertaken by IA are given below:

End User : British Steel, Scunthorpe
Shell Used: Crystal
Reference : (15)

Purpose

The Knowledge Based system at Scunthorpe is used to monitor the waste gas recovery from the Steel Making Process, the gas is subsequently used for fuel. The system is capable of real-time monitoring of the gas collection system and of giving rapid detection of faults.

End User : Exxon Chemicals
Shell Used: Crystal
Reference : (16,pers. comm.)

Purpose

This is an on-line continuous monitoring Knowledge Based system which analyses vibration characteristics from the main pumps and compressors within the plant. The system is capable of detecting early signs of impending mechanical failure. Another on-line Knowledge Based system is currently being developed to interpret data from the primary gas turbine.

End User : British Steel, Ravenscraig
Shell Used: Nexpert
Reference : (15,17)

Purpose

The system at Ravenscraig is a large VAX based system to monitor and identify faults on the entire Steel Making Process. The system is used by British Steel's Maintenance Engineering staff. The system represents one of the largest Knowledge Based systems projects ever installed in the UK.

Another Knowledge Based system tool is the PROMASS expert system shell which was jointly developed by British Nuclear Fuels plc (BNFL) and Unipro Ltd (18). At present one application of the system is to monitor the Integrated Dry Route process for the conversion of Uranium Hexafluoride to Uranium Dioxide. The main application of the PROMASS system is in the monitoring of the Vitrification plant at BNFL's Sellafield plant. At Sellafield the Vitrification plant converts high level radioactive liquid waste into a glass product for long term storage. The PROMASS system collects data from 2500 instruments and interprets the signals via 10 000 rules. The PROMASS system is written in C and is currently being marketed as a commercial process monitoring and control system.

At the present time several of the above Knowledge Based system shells are being evaluated, although no decision has yet been reached.

In addition to acting as a closed loop control and monitoring system a consultative Knowledge Based system is envisaged attached to the main system. The consultative system will be used by the operator to advise on routine maintenance, and give condition monitoring capabilities from data gathered manually. In the case of vibration analysis a system of this type would advise on bearing condition after the manual input of data collected by the operator. Also incorporated into the consultative side of the Knowledge Based system will be the control algorithm for the crusher. The control algorithm is being jointly developed between Pegson Ltd. and Leicester Polytechnic. The algorithm will be based around the research of Bearman, (as previously described) whereby the power and product grading can be predicted given the mechanical properties of the rock and certain controlling parameters associated with the crusher. Data input through the consultative Knowledge Based system will generate initial conditions which will be available to the main control system. If at any time the material properties of the feed change this can be entered through the consultative system which will then adjust the control algorithm.

A further development in the field of Artificial Intelligence is the emergence of neural networks. Neural networks effectively model the thought processes of the human brain. Both the artificial neural network and the biological equivalent consist of a large number of simple elements that learn and are collectively capable of solving problems. A hybrid Knowledge Based system/neural network may represent the best control strategy and at present this is being investigated.

4 CONCLUSIONS

The control system proposed represents a mechatronic approach to the development of an on-line control and monitoring system for cone crushers fitted with hydraulic setting adjustment. The initial control algorithm is generated via a consultative expert system knowledge base which is linked to the main real-time control and monitoring knowledge base. The control and monitoring system receives data from an array of sensors, the data then being analysed via a set of rules. As a result of the analysis actions are advised and automatically implemented by the system. Data can be examined on-line or stored in a database from where historical trending can then be carried out at a latter date. On-line fault diagnosis is another feature of the system which advises on possible failures and routine maintenance. Hypertext is to be used in the fault diagnostics package to provide in-depth information to the level of the traditional parts and service manual. Another consultative knowledge base is available to respond to manual input of information regarding data collected independently of the integral sensors. It is expected that data pertaining to vibration and bearing condition may be included in this knowledge base.

The potential increase in efficiency offered to the quarrying industry in the U.K, in terms of reduced downtime, improved product and improved utilization of energy is substantial.

ACKNOWLEDGEMENTS

The work outlined in this paper is being supported by the Department of Trade and Industry via their Support For Products Under Research (SPUR) scheme.

The authors would like to express their thanks to Mr C. Hall, Senior Development Engineer at Pegson Limited for his assistance during the preparation of this paper.

REFERENCES

1. Bearman R.A., Energy utilization in quarrying, Mine and Quarry, September 1991.
2. Bearman R.A., Barley R.W. and Hitchcock A. Prediction of power consumption and product size in cone crushing, Minerals Engineering, Vol. 4, No. 12, pp 1243-1256, Pergamon Press, 1991.
3. Barton C.C., Variables in fracture energy and toughness testing of rock, Proc. 23rd US Sym. on Rock Mechanics, pp 449-462, Ed: Goodman R.E. and Heuze F.E., AIME, New York, 1982.
4. Sun Z. and Ouchterlony F., Fracture toughness of round specimens of Stripa Granite, Int. J. Rock Mechanics Min. Sci. and Geomech. Abs., Vol 23, pp399-409, 1986.
5. International Society for Rock Mechanics (ISRM), Suggested methods for determining the fracture toughness of rock material, Co-ordinator: Ouchterlony F., 1987.
6. Harris C.A. and Kosick G.A., Expert System technology at the Polaris Mine, Paper No. 9, 20th Annual Conference of the Canadian Mineral Processors, Ottawa, January 1988.
7. Meech J.A., Expert systems for teaching and training in the minerals industry, Minerals Engineering, Vol. 3, No. 1/2, Proc. of the International Comminution Sym., Pergamon Press, (1990).
8. Harris C.A., Comdale Technical Literature, (1991).
9. Lacouture B., Russell C., Griffin P. and Leung K., Copper flotation expert system at Mount Isa Mines Limited., Presented at Copper '91, Ottawa, Canada, August 1991, (1991).
10. McDermott K., Cleye P., Hall M. and Harris C.A., Expert system for control of No. 4 autogenous mill circuit at Wabush Mines., Presented at the Canadian Mineral Processors and Operators Conference, Ottawa, Canada, January 1992, (1992).
11. Harris C.A., Sprentz P. Hall M. and Meech J.A., Improving Productivity: Justification for the use of expert systems in the process industries. Comdale Technologies Inc., May 1990, (1990).
12. Spring R. and Edwards R., Real-time expert system control of the Brenda mines grinding circuit., AIME Pre-print No. 89-58, Presented at SME Annual Meeting, Las Vegas, Nevada, February 27th-March 2nd 1989.
13. Anon., High-tech operations - the manless mill, Northern Miner Mag., Vol. 5, No. 5, May 1990, pp 22-23.
14. Dept. of Trade and Industry publication, Real-time process control: Improved efficiency - Blue Circle Industries plc (LINKman application), HMSO, (1990).
15. Milne R., Case studies in condition monitoring., Knowledge-based Systems for Process Control, Ed: Grimble M.J., McGhee J. and Mowforth P. Pub: Peter Peregrinus Ltd., pp255-66, (1990).
16. Milne R., Personal Communication, Intelligent Applications, Current Project List, (1992).
17. Milne R., Bain E. and Drummond M., Predicting faults with real-time diagnosis., IEE, International Control Conference, Brighton, England, December 1991.
18. Harper W.J., The development of expert systems for process monitoring in British Nuclear Fuels, BNFL, (1990).

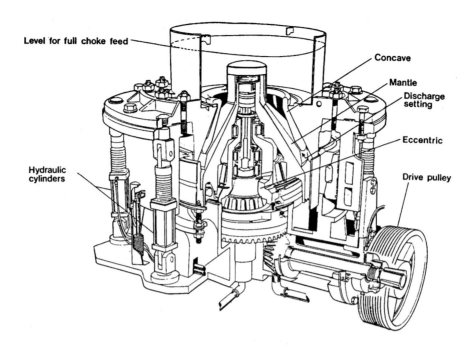

Fig 1　Pegson Autocone Crusher

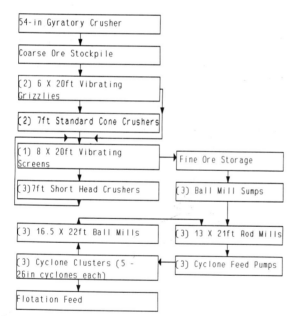

Fig 2　Typical mineral processing circuit

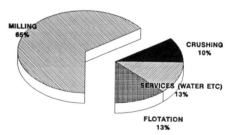

Typical Copper-molybdenum deposit

Fig 3　Mineral processing energy analysis

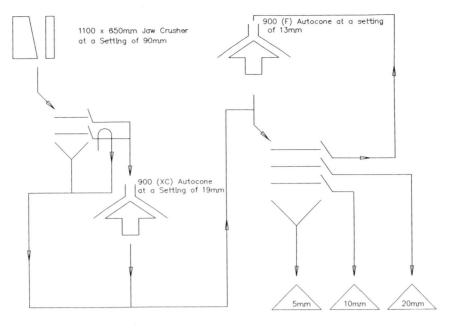

Fig 4 Typical quarry flowchart

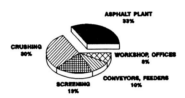

Fig 5 UK energy consumption

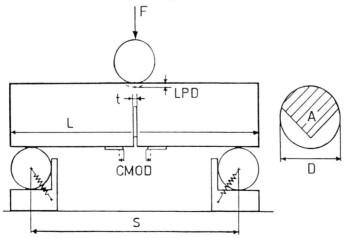

Fig 6 Chevron bend test

An integrated mechatronic research cell for the decoration and assembly of scale models

L G TRABASSO, MSc, PhD, A P SLADE, J R HEWIT, FIMechE, MIEE, CEng
Loughborough University of Technology, UK

SYNOPSIS As a direct result of the work undertaken for an SERC supported research programme the authors have developed a modular printing/assembly cell, which they believe can also be applied to other industries.

1 INTRODUCTION

Mechatronic design principles have been applied to the development of a multi robot research cell for the decoration and assembly of scale models.

The software consists of three major packages named EAGLE, OWL and PUFFIN. There are also routines which drive a modified Tampo printing machine and a conveyor system.

EAGLE is a VAL II programme which at present acts as the overall cell controller. It is used to control the operation of a PUMA robot and to distribute the information concerning the status of the cell. **OWL** is a Microsoft Pascal programme which performs identification/inspection operations using visual data and passes relevant measurements and pass/fail parameters to the appropriate machinery. **PUFFIN** is written in FRTX Forth and is the operating programme for an RTX robot.

Various controls of a Tampo printing machine including those for cycle operations, manual intervention and auxiliary functions have been parallelled in software and are accessed via a separate VAL II programme called **TUCANO** under the control of EAGLE.

The initialisation and start-up procedures for the cell are being implemented in the expert system shell Savoir which will employ sensory feed back in identified critical areas to indicate the operational status of the cell, and this will eventually be integrated into the software packages.

The integration of this software with the Mechatronically designed handling and decorating machinery is described. This integrated design study has led to concepts for a new generation of pad printing machinery which will increase speed, precision, reliability and flexibility.

2 OVERVIEW OF THE DECORATION CELL.

The decoration cell currently consists of five major stations: *1) loading, 2) identification, 3) decorating, 4) inspection, 5) unloading* built around a conveyor system. The sixth component is an overall cell controller which is at present being developed. This will take the operator through the start-up procedures and give information about incorrect settings or choices. It is also intended that this controller will also be used to take the operator through the teaching procedure with new models to optimise and speed up operation.

The third function of the controller will be the constant monitoring of the cell during operation and to signal faults and bring the cell to a halt in an orderly fashion in the event of a fault.

A major obstacle to operational integration is the number of high level languages used to control the cell constituent machines as shown in Table 1.

These various languages are standard features of the various components of the cell; VAL ll with the PUMA controller, MS Pascal with the JOYCE-LOEBL MAGISCAN vision system and FRTX Forth with the RTX controller.

Table 1: The various high level languages used in the decoration cell.

Loading	PLC mnemonic code
Identification	MS-Pascal and VAL ll
Decoration	VAL ll
Inspection	MS-Pascal and VAL ll
Unloading	Forth and VAL ll

3 LOADING STATION.

The loading station consists of a chute and a simple mechanism with two degrees of freedom, shown in Figure 1.

Fig 1 The loading station

The loading station is under the control of a PLC programme called TICO-TICO. This is the main PLC programme used to move the pallets around the conveyor system and communicate with the peripheral devices. The PLC is programmed via a set of 22 basic instructions in its own mnemonic code(1). Examples are **LD** (LoaD), **LDI** (LoaD Inverse), **AND** (AND), **OUT** (OUTput), **S** (Set) and so on. The mnemonic code is best expressed as a ladder diagram and a sample is given in Figure 2.

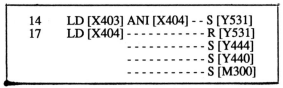

Fig 2   Part of the mnemonic code of TICO-TICO

In this sample of code the pallet can only enter the loading station when it is empty (ANI X404). When the pallet arrives (LD X404) the next pallet is locked out (R Y531) and the sequential action of the loading station commences.

The loading station does not interact with the main control programmes of the cell.

4   IDENTIFICATION STATION.

This section presents the software used by the vision system and the PUMA robot-vision system interface. The software is split into two packages: off-line and on-line called **OWLDB** (OWLDataBase) and **OWL** respectively. The off-line package is used in the generation of the pattern recognition and the vision-PUMA calibration databases; these are then used by the on-line software for identifying the bodyshells and managing the robot-vision interface.

To make the software user friendly the generation of the inspection database was also included in OWLBD. The operator can therefore generate all the databases needed through a single programme. The structure of OWLDB is represented in Figure 3.

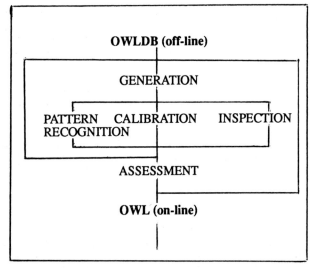

Fig 3   Structure of the off-line programme of the identification station

The data generated by the off-line programme OWLDB, namely the pattern recognition and calibration databases, is used by the on-line programme in order to:

- identify the bodyshell on the pallet

- drive the PUMA robot to pick up the bodyshell from the pallet.

To identify a bodyshell the vision system takes a set of measurements and compares it with those held in the database, and based upon the distance from the set in feature space, will either accept or reject it. There is a tolerance that has to be set in the classification process in order for the bodyshell to be accepted as a member of the database. This tolerance level is defined as the <u>confidence level</u> of the process and is set by the user. The minimum and maximum levels in the MAGISCAN system are 0 and 9 respectively.

After successfully identifying a bodyshell the vision system sends the following information to the PUMA robot:

- Bin number, which uniquely identifies each model in the library

- The **x** and **y** co-ordinates of the centre of boundary of the bodyshell, already transformed into robot world co-ordinates by the factors in the calibration database

- Orientation angle, which is measured from the reference model in the database.

With this information the PUMA robot programme EAGLE modifies the location of the centre of boundary of the bodyshell as follows:

SET<bodyshell> = TRANS (x,y,z,o,a,t)

where **x,y** and **o** are sent by the vision system, **z, a** and **t** are set by the operator: **z** is the height of the bodyshell measured from the top of the pallet, **a** and **t** are the angles which determine the relative orientation of the robot gripper.

5   DECORATION STATION.

The software of this station consists of three sub-routines in **TUCANO** which take control of the Tampo printing machine. They are:

SWEEP.TAMPO - continuously keeps the doctor blade/spatula carrier sweeping ink over the cliché when the PUMA robot is not at the Tampo printing machine.

STOP.TAMPO - stops the Tampo printing machine so allowing the PUMA robot to manoeuvre the matrix in front of the printing machine.

PRINT.TAMPO - executes the preset print cycle of the printing machine.

These routines are very simple, consisting mainly of software to switch the appropriate output lines on the PUMA I/O board. Figure 4 lists the sub-routine **TUCANO** from where these routines are called.

```
1   SPEED 80 ALWAYS      ;set speed to medium
2   MOVE #bonnet.out     ;move robot in front
                         ;of printing m/c
3   WAIT TIMER (-1) == 0 ;wait for robot to
                         ;get to location
4   CALL stop.tampo
5   WAIT SIG(2001)       ;printing m/c stopped
6   DISABLE CP           ;stop software
                         ;'rounding' corners
```

```
7     MOVES b1up           ;move above first
                           ;print location
8     MOVES b1             ;move to first print
                           ;location
9     WAIT TIMER (-1) == 0;wait for robot to
                           ;get to location
10    CALL print.tampo
11    WAIT SIG(2001)       ;finished printing
12    MOVES b2up           ;move up from final
                           ;print position
13    CALL sweep.tampo     ;restart 'idle'
                           ;cycle
14    flag = TRUE          ;needed to keep
                           ;sweep.tampo running
15    DEPARTS 250          ;move away from
                           ;printing machine
16    ENABLE CP            ;go back to normal
                           ;mode
17    MOVE rupb            ;move round from
                           ;printing machine
18    SPEED 100 ALWAYS;set speed to fast
19    RETURN
```

Fig 4   The VAL II code of TUCANO

6   INSPECTION STATION.

The vision system inspects the decorated bodyshells for:

position and completeness of the decoration

- smudging

- shades of colours

The inspection is based upon a comparison of the decorated bodyshell with that of a reference model. All the relevant features of the reference model are derived off-line and stored in a database. The on-line inspection programme reads in this database and compares the features of the decorated model with that in the database.

This process is similar to pattern recognition although it is simpler in that the correlation of the features is not relevant, i.e. if the position of the decoration is incorrect then this gives the inspection software a basis for a decision; it is clearly not necessary to investigate further, however.

The programmes were developed to include a feature which makes the decoration inspection very quick and reliable windows allocation (2).

The principle of windows allocation is very simple: the off-line programme allows the user to define windows around the features to be inspected, and the origin and size of the various windows are recorded in the inspection database. The on-line programme rebuilds these windows and then scans the first one. If the relevant decoration is found and its attributes are within the preset tolerances the next window is scanned and so on, otherwise the bodyshell is rejected at the first window.

This windows allocation approach allows for a tremendous reduction in the number of pixels scanned. In the bonnet inspection of the Texaco Sierra, for example, the number of pixels scanned is reduced by 97% from 64,536 to 2,000 pixels.

The movement of the PUMA is controlled by EAGLE, and the vision software OWL sends data to PUFFIN via EAGLE. At present this is simply pass/fail but eventually the reason for failures will be sent back so that full quality control will be possible.

7   UNLOADING STATION.

PUFFIN is the programme which controls the RTX robot, it is written in **Forth** and structured in elementary units called **words**. A word in Forth corresponds to a procedure in Pascal or a sub-routine in VAL II. With reference to Figure 5, it is easy to follow the actions performed by the RTX robot just by looking at the **PUFFIN** source code.

```
0     :PUFFIN{MAIN PROGRAMME}
1        CLEAR_SCREEN  0 COUNT !  0 DECISION !
         0 MODELS !
2        ATTACH_GRIPPER
3     BEGIN
4        REC DECISION !
5        DECISION @ 84  = IF FIRST_TRAY
6     ELSE DECISION @ 66 = IF  STORE_TRAY
7     ELSE DECISION @ 80 = IF PALLETISE
8     ELSE DECISION @ 102 = IF REJECT
9        THEN THEN THEN THEN
10       DECISION @ 83 = UNTIL
11    CR  ." Finishing"  CR  HOME  CR
         . " Finished "  CR ;
```

Fig 5   The PUFFIN source code

The word **REC** receives action data from the PUMA robot via the RS-232 link and stores them in the variable **DECISION**. Based upon the current value of this variable the RTX robot carries out one of the following actions:

DECISION = 84: (ASCII code T for tray) the RTX robot moves the top transport tray from the 'empty' stack to next position on the 'load' stack where the decorated bodyshells which pass inspection will be palletised

DECISION = 66: (B for bin number) after identifying a bodyshell at the identification station the vision system sends the RTX controller the corresponding bin number. The word **STORE_DATA** stores this variable in the array **MODELS**.

DECISION = 80: (P for palletise) the bodyshell has passed inspection and the RTX palletises the bodyshell. The word **PALLETISE** is composed of the longest chain of words within **PUFFIN** as it involves interaction with the PUMA robot and the PLC as well the updating of a number of arrays and variables.

DECISION = 102: (f for fail) this tells the RTX controller to place the next bodyshell in the reject/rework box.

DECISION = 83: (S for stop) This is the stopping code. The PUMA robot controller sends the same signal to the vision system and the PLC as well. After receiving this code the RTX robot returns to its **HOME** position.

The main action of the RTX robot is the palletisation of the correctly decorated bodyshells. An overview of the corresponding code follows. The initial action of the word **PALLETISE** is to read the first value of the array **MODELS**. This will decide which bodyshell will be palletised and onto which transport tray. The array **MODELS** is updated and the code moves down to the next word. Based on the value of a counter this word decides on which of the six possible locations on the tray the bodyshell will be placed.

Even though there are 30 palletising positions only 6 needed to be taught to the RTX as the co-ordinates for each tray differ only in height. When the value of the counter reaches six the RTX robot places a new tray on the 'load' stack, taking into account the number of trays already on the stack.

An example of the code needed to palletise a Ford Sierra at the first position of a transport tray is given in Figure 6. **ABOVE_PUMA** and **FROM_PUMA** are taught locations. Note that the RTX controls the vacuum ON and OFF in lines 1 and 7 with the words SUCK_MODEL and BLOW_MODEL respectively. The command **BIT** is used to control the output port linked to the PLC which switches the solenoid valves linked to the suction cups ON and OFF accordingly.

The word **ROBOT_TRAFFIC** is used for synchronising the movements of the robots. **PC!** is a FORTH command that sends data over the RS-232 link to the PUMA controller, it is used in this instance to signal that the bodyshell has been grasped and that the PUMA robot can now be moved away from the unloading location. When the PUMA robot has moved away a signal is sent back to the RTX controller. Note that in line 2 of the word **ROBOT_TRAFFIC** that the RTX robot will not move until it receives this signal, this is achieved by the use of the FORTH statement **PC@**. Line 3 is the handshake code for the PUMA-RTX serial link.

```
0  : SIERRA[1]   {model location}
1     ABOVE_PUMA  FROM_PUMA  SUCK_MODEL
2     ROBOT_TRAFFIC
3     GET_HEIGHT_UP_MODEL
4     -1544 312 RITE_Z -756 -1636 -489 0 GOTO>POS
5     GET_HEIGHT_MODEL
6     -1544 312 RITE_Z -756 -1636 -489 0 GOTO>POS
7     BLOW_MODEL
8     GET_HEIGHT_UP_MODEL
9     -1544 312 RITE_Z -756 -1636 -489 0 GOTO>POS
10    100 ZED !SPEED ABOVE_PUMA 160 ZED !SPEED ;
```

Words used in the above code

```
: SUCK_MODEL  1 0 0 ! BIT ;
```

```
0  : ROBOT_TRAFFIC
1    71 760 PC! {model grasped -> move on}
2    BEGIN 760 PC@ 77 = UNTIL {from Puma }
3    {path clear -> move on}
4    42 760 PC! {to Puma : over } ;
```

```
: GET_HEIGHT_MODEL
    MODEL_HEIGHT @ ZED !NEW_POS ;
```

```
: WRITE_Z  ZED @ NEW_POS ;
```

```
: BLOW_MODEL  0 0 0 ! BIT ;
```

Fig 6 An example of the palletising code

8 THE STARTUP PROCEDURES.

It was found that some artificial intelligence tools could be applied to the robotic decoration process to improve performance at certain stages of its operation. There are presently two such tools under development: a) start up of the cell and b) assessment of the training process. Both applications are implemented using the expert system shell Savoir (3).

8.1 Start up of the cell.

This tool consists of a package to aid the operator to carry out this task, which consists of the calibration of the robots, establishment of the communication between the several controllers, setting the correct air pressure for the robot tools and the conveyor system, setting the correct voltage for the lighting system and so on. The more this operation is automated the easier it will be for the operator and to this end sensors are being fitted to various parts of the cell to measure important parameters.

8.2 Assessment of the training process.

It can easily be shown that a poor training process will lead to poor results in the classification process. The consequences of this can be disastrous, since the robot is driven to pick up the model based upon the results of the classification process.

Unfortunately, the result of the training process can be different when executed by different operators, since some decisions are left to the operator's judgment; the definition of the threshold levels to transform a grey image into a binary image is just one example. Based only on visual inspection, the operator has to decide whether the binary image obtained is 'good enough' to represent the model in view.

The approach presently adopted to tackle this problem is to leave the training process with the same degree of subjectivity that it has always had and to use the knowledge stored in an expert system to 'judge' the result of the training process based on information gathered from the classification process.

9 REFERENCES.

(1) Mitsubishi Electric Corporation. Melsec Programmable Controller Programming Manual. 1987.

(2) L.G. Trabasso, A.P. Slade, J.R. Hewit. Intelligent Decoration of Scale Models. Fourth World Conference on Robotics Research, Pittsburgh, Pennsylvania, USA. 1991 (SME)

(3) Intelligent Systems International. The SAVOIR expert system shell: programmers guide. Surrey, UK, 1987.

The Lancaster University Computerised Intelligent Excavator programme

D A BRADLEY, BTech, PhD, MIEE, **D W SEWARD,** BSc, MSc, MICE
Lancaster University, UK

SYNOPSIS The Lancaster University Computerised Intelligent Excavator (LUCIE) programme is aimed at achieving fully autonomous excavation over a wide range of ground types and conditions. Based upon the use of a one-fifth scale model of a back-hoe loader arm together with an implementation on a JCB801 tracked excavator, the LUCIE system is currently capable of independent operation to excavate a rectangular trench to a preset depth in a variety of soils. The system is truly mechatronic in nature incorporating the servocontrol of the actuators governing the movement of the excavator arm, the associated sensors and software necessary for both tactical and strategic decision making.

1 INTRODUCTION

The civil and construction industries currently deploy large numbers of manually controlled and operated plant for a wide variety of tasks within the construction process with, in general, the quality of the result being a function of the skill of the driver/operator acting in collaboration with other members of the site workforce. The introduction of the technologies of automation and robotics into what currently remain primarily manual industries is predicted as having a major impact on the way in which these industries operate in a manner similar to that experienced in the automation of the more conventional manufacturing industries [1,2].

There are, however, major differences between the ways in which automation and robotics has been and can be applied in the two industries. In manufacturing, the introduction of automation and robotics has, in general, been accompanied by a structuring of their operating environment to meet the needs of individual items of equipment. In contrast, a construction site presents a complex, and indeed, hostile environment in which the plant must function and survive. This in turn requires that construction robots must be provided with control structures and hierarchies which will enable them to function effectively in a site environment and implies the integration of mechanical function with advanced software structures and a complex sensor environment which characterises a mechatronic system. Indeed, robotic construction plant may be considered as representative of a class of 'large mechatronic' systems which have as their primary function the provision of high forces at long reach and incorporate advanced features for self management and control.

Earth removal in association with operations such as trenching is a common feature of a construction sites and is one which is usually carried out by means of either a general purpose or dedicated item of plant under manual control with assistance provided by one or more additional members of the workforce. The nature of the task and of the working environment means that performance is to a very significant degree dependent upon the skill of the operator. The introduction of robotic plant for this tasks has been shown to have advantages in a number of areas, particularly in respect of quality and the reduction of hazard to the operator [3].

The Lancaster University Computerised Intelligent Excavator (LUCIE) programme to be described is aimed at deploying mechatronic principles in achieving an autonomous robot excavation capable of functioning over a wide range of ground types and conditions. Based on the use of a one-fifth scale model of a back-hoe loader arm in the laboratory which serves as a development tool for hardware and software concepts together with an implementation on a JCB801 360 tracked excavator the LUCIE system is currently capable of independent operation to excavate a rectangular trench to a preset depth in a variety of soils and can deal with and remove a range of obstacles from within the trench. The system is truly mechatronic in nature incorporating the servocontrol of the actuators governing the movement of the excavator arm, the associated sensors and sensing systems and the high level software structures necessary for both tactical and strategic decision making in the excavation process.

2 PROBLEM DEFINITION

In operation, the primary objective of an excavator may be summarised as:

"Fill the bucket with earth as rapidly as possible"

and then to repeat this as often as is necessary

within the constraints imposed by the characteristics of the trench being dug until the required trench profile is achieved. Quality may be assessed in terms of the finished trench taking into account factors such as the absence of any overdig and the subsequent need to backfill the trench, an expensive and time consuming procedure.

A more detailed examination of the sequence of events associated with one operational cycle - bucket empty to bucket empty - is shown by figure 1 from which it can be seen that the task structure can be separated into those associated with the movement of the bucket in air and those associated with the movement of the bucket through the ground. Each of these motions presents particular problems and can be considered individually.

2.1 Movement in air

On initial inspection, the movement of the excavator arm in air may be considered as equivalent to the motion of a conventional anthropomorphic robot and that similar control and motion strategies can be deployed including time co-ordinated and time unco-ordinated joint motions, point-to-point motions and path following. However, a fuller inspection of the problem indicates that this is not necessarily the case. In particular, the hydraulic system supplying the actuators for the excavator arm is unlikely to be capable of providing a flow capable of supplying the combined maximum demands of all the actuators and this will constrain the nature of the motion that can be achieved.

Consider in more detail the motion of the bucket after emptying. Based on a knowledge of the previous pass through the ground and the required trench profile the bucket must be moved as rapidly as possible to an appropriate starting point relative to the trench and, in doing so, must avoid any possibility of collision with obstacles on its path. Once in position in relation to the trench, the bucket must be brought into contact with the ground in order to start the digging process.

Once the bucket has been filled it must then be removed from its position in the trench and moved to the point at which it is to be emptied, which may well vary with each operating cycle. During this phase the bucket is likely to be carrying a load which is significant in terms of the mass of the arm, unlike a typical assembly robot where the robot structure is quite massive in relation to its payload. The resulting stresses on the arm and the constraints on the acceleration and deceleration profile that can be adopted are therefore much more severe than would be encountered with a conventional robot.

Motion of the arm in air must therefore be associate with real-time path planning routines which incorporate some form of knowledge about the operating environment, the position of obstacles in that environments and the nature and behaviour of the structure of the arm.

2.2 Motion in ground

Once in the ground the bucket must follow a path which is designed to fill it as rapidly as possible while, at the same time, allowing it to deal with a variety of ground conditions. In particular, the controller must be able to modify, in real-time, both the digging strategy and the associated tactics that are to be adopted. Of especial importance during this phase of operation is the ability of the controller to adapt to changes in the ground conditions to deal with the presence of unexpected obstacles, such as boulders and rocks, on the prospective path of the bucket. It is also important that the system should be aware of the presence in its path of services such as power cables, communications cables, water mains and sewers and take action to avoid damaging any of these.

3 THE LUCIE SYSTEM

The LUCIE system is currently implemented on the JCB801 360 tracked excavator shown in figure 2. This retains the basic arm structure, actuators, motor and hydraulic supply onto which the system has been built. Additional features include the sensor fit, a set of proportional hydraulic valves operating under computer control in parallel with the existing manually controlled valves and the on-board computer. This leads to the configuration shown in figure 3 in which the selection of the automated and robotic environment or the original, manual system is by means of a changeover valve in the hydraulic supply. At this stage in development, the manual system for operation of the arm has been retained in order to facilitate recovery in case of any failure in the automated systems. Also, LUCIE is not intended to be self mobile and the tracks are therefore under manual control only.

Consider the general schematic of a mechatronic system shown as figure 4, each of the major elements of which - sensors, processor, software, operator interface and actuators - are incorporated in the LUCIE systems and can be considered separately and in detail.

3.1 Sensors

Once in position, the body of the vehicle deploying the excavator arm becomes the reference for the motion of that arm. The initial sensor fit is therefore concerned with the provision of information on the position of the tip of the bucket with reference to the attachment point on the vehicle body.

In the case of the one-fifth scale laboratory model used for software development [4], plastic track rotary potentiometers were used to measure joint angles. Though effective and reliable in the context of the model the use of potentiometers was not considered to be a practical option for use on the 801. Instead, absolute encoders were mounted in specially designed housings on each of the 'shoulder' (boom), 'elbow' (dipper) and 'wrist' (bucket) joints of the arm. Experience has shown these encoders and the housings to be robust and reliable, even when operating the excavator in muddy soils or even underwater.

As has already been stated, the body of the vehicle provides the reference for the movement of the arm. In operation it is possible under certain conditions, for example on encountering an obstruction on the bucket path for the motion of the actuators to be translated not into a movement of the bucket through the ground but into a moment which rotates the body of the vehicle, lifting the front off the ground as in figure 5. In order to detect this event a tilt sensor is mounted in the cab of the vehicle aligned with the digging plane.

Finally, cab rotation is monitored using the slew gear as a reference for two magnetic sensors placed at a quarter of a tooth pitch together with a third sensor to provide a zero.

3.2 Hydraulic System and Actuators

Both the one-fifth scale laboratory model and the JCB801 are actuated by double-acting hydraulic rams, the oil supply to which is controlled by a set of linear proportional electro-hydraulic valves operating under software control. It is ultimately intended that these relatively expensive valves are replaced by simpler valves incorporating local control in line with the general mechatronic philosophy of replacing mechanical complexity with local intelligence as part of a distributed system.

The hydraulic system is supplied with oil by pumps driven by the on-board diesel engine. As has already been noted, the engine and pumps do not have the capacity to supply the maximum demand of each actuator simultaneously and, for economic reasons, it is envisaged that this situation will persist even as the degree of system automation increases.

3.3 Computer System

Figure 6 shows a block diagram of the electronic system. The processor used is the Harris RTX2000 FORTH engine. This is designed specifically for and around the FORTH language and is estimated to be some 40 times faster than a 10MHz Motorola 68000 processor for real-time applications. The RTX2000 achieves this level of performance by virtue of the fact that its design and implementation in silicon allows an instruction to be fetched, decoded, executed and the address of the next instruction to be placed onto the bus in the course of a single clock cycle, even in the case of a jump or conditional branching instruction. Additionally, the internal architecture comprises 4 independent buses addressing external memory, I/O and the two internal stacks respectively. These buses are controlled by separate bit-fields in the instruction and can be used concurrently. The result is that with the use of an optimising compiler up to 5 FORTH words can be compressed into a single 16-bit instruction in the RTX.

All the electronic hardware is currently mounted on standard Eurocards and connected via an STE (IEEE 1000) bus and is contained in a weatherproof housing mounted on the roof of the 801. External connections are generally by means of military standard weatherproof plugs and sockets. Consideration will be given in the next stage of development to the design and production of application specific integrated circuits (ASICs) to take over man of the roles currently carried out by means of discrete cards.

3.4 Software Structures

As indicated, the software for LUCIE is written in FORTH and covers both the servocontrol functions associated with the movement of the arm and the higher level AI type functions associated with decision making processes at both the tactical and strategic levels. These managerial functions include tasks such as:

- Task planning and scheduling
- Safety
- Process checking
- Monitoring and reporting
- Task updating

To meet these requirements, the control hierarchy of figure 7 is proposed. At the lowest level of this hierarchy is the servocontroller responsible for the control of arm motion. This will, in general, operate under the direction of the next level in the hierarchy which has responsibility for the real-time strategic and tactical decisions associated with filling the bucket and dealing with obstructions. This level is referred to in the LUCIE system as the 'Activities Manager' and receives information from both the servocontroller and the sensor environment. The Activities Manager would also assume responsibility for local managerial functions.

The higher levels in the control hierarchy would be concerned with general supervision and would be responsible for the direction and co-ordination of a number of robot sub-systems. At this supervisory level information relating to the general planning and scheduling is available, including access to a full range of on and off-site systems in order to provide both strategical and tactical advice and support on request.

The Activities Manager - The software for the Activities Manager and servocontroller has the major subdivisions as shown in figure 8. The Activities Manager itself is a high-level, rule-based controller which is the source of intelligence within the LUCIE system. It currently contains around 70 rules, some of which have been generated as a result of observation of skilled operators over a variety of tasks. The program structure is that of a production system utilising the RTX processor and the FORTH language by means of the FORPS structures [5].

The production system rule memory contains all the rules in a form such as:

IF (condition is true) THEN DO (action)

A typical rule would be:

IF (penetration of bucket > 300 mm) THEN DO
(rotate bucket)

The use of adaptive rules means that it is impossible to predict the path of the bucket through the ground as the digging strategy and tactics are continually being adapted by the Activities Manager to meet changing ground conditions.

The working memory for the system contains all the current values of the variables used for checking the conditional element of the rules. The values of these variables can be altered either by the servocontroller, by the production rules or by the system sensors. This part of the memory can be considered as a 'blackboard' on which the latest information about systems status is presented for use by any part of the program.

The structure of the program is such as to facilitate the addition of extra rules without changing the basic program structure. In the case of trenching the requirements are largely procedural and the production rule memory has therefore been partitioned so that at any one time the inference engine only checks the relevant rules rather than the full set.

Operator Interface - At present the on-board operator interface consists of a thumb-wheel type controller for setting the depth of the trench relative to the vehicle tracks.

The software interface is provided by means of a conventional PC which is used as the development vehicle for the FORTH structures prior to downloading to the RTX2000 on LUCIE. However, it must be noted that the FORTH language does not produce easily readable programs and it is difficult for non-expert programmers to add to or modify the rule structure. Work is therefore underway to develop a programming environment which enables rule to be written in a more natural form before conversion to the Reverse Polish Logic notation used by FORTH. This would enable rules to be generated in a form such as:

"If the bucket cannot move fast enough through the ground then raise the bucket slightly and try again"

The development of this 'dig language' is seen as the mechanism by which the basic functionality of the system would be modified on-site to meet specific needs.

4 FIELD TRIALS

LUCIE has now been field tested many times in a wide variety of ground conditions and types and the rules modified and tuned to enable effective digging, achieving a full bucket in virtually every dig cycle. In the process of testing and trials, the system has demonstrated its ability to cope unaided with large obstructions such as boulders and to autonomously develop a strategy to remove the obstacle from the line of the trench. Depth accuracy is well within the target of and a high quality, flat-bottomed trench results which is of a standard comparable to that achieved by a skilled operator and better than that of an average or novice operator.

The cycle time - from emptying bucket to emptying bucket - is currently of the order of 30 seconds which is about double that which can be achieved by a skilled operator. Work is currently underway to improve this to the point where cycle times comparable with those of a skilled operator are achieved.

5 FUTURE DEVELOPMENT

The project has achieved its initial aim of autonomously digging a trench to a defined depth in a variety of ground conditions and types and may be considered to have proved the concept. In the short term, many of the techniques already developed could be incorporated into excavator design as the basis of operator support systems to simplify the operators task and to provide features such as the prevention of overdig. It also becomes possible to consider new geometries for the excavator arm, perhaps incorporating additional degrees of freedom. In this case the operator will control the motion of the bucket, for instance by the use of a multi-degree-of-freedom joystick, with individual joint motions coming under direct computer control.

However, much further development is necessary for both the on-board servocontroller and Activities Manager before a truly autonomous system could be produced. Of particular importance are the provision of an improved programming technique and associated high-level programming structures using natural language forms to simplify the entry of rules and to form the basis for the user interface with the controller - the 'dig language'. Other areas for investigation and research include the sensor environment and the increased integration of sensors into the control hierarchy.

In addition systems such as those required for safety, mapping the local operating environment, precise positioning and location, including the monitoring of the 6-degrees of freedom of the mobile base, the detection of underground obstructions such as cables and pipes and the linking to on and off-site systems require investigation and research before a truly autonomous system can be produced. Many of these topics are the subject of current research programmes in the Construction Robotics Group in the Engineering Department at Lancaster.

6 CONCLUSIONS

Unlike conventional manufacturing, the operating environment of a construction site can only to a limited degree be structured to accommodate automated and robotic systems and such systems must therefore be structured to enable to function and survive in a site environment. This requires the development of controllers and control systems which are significantly more flexible and adaptable than those currently used for most robot control, including a real-time decision making capability at both the tactical and strategic levels.

The combination of the one-fifth scale laboratory model and implementation on the JCB801 has provided a powerful and effective research environment for investigations into construction robotics and have demonstrated the effectiveness of an AI approach to the control and operation of robot systems in complex environments in the form of the LUCIE system. Using this approach, the ability to dig, without operator intervention, a high quality trench has been demonstrated in a variety of ground types and conditions, including the ability to deal with unknown obstacles within the trench profile.

Further, many of the software techniques developed for LUCIE are considered to have implications for other types of construction robots, particularly where applications demand a subtle behaviour capable of being described by rules.

References

1 Hasegawa, J., 1991, International Association for Automation and Robotics in Construction, Newsletter No 1, June

2 Seward, D.W., Bradley, D.A. & Garas, F.K.,1991, 'Towards Site 2000', 8th International Symposium on Automation and Robotics in Construction, Stuttgart, Vol 1, pp131-137

3 Bradley, D.A. & Seward, D.W., 1990, 'Robots and Automated Systems for the Civil and Construction Industries', Civil Engineering Systems, Vol 7, No 3, pp135-139

4 Seward, D.W. & Bradley, D.A., 1988, ' 'The Development of Research Models for Automatic Excavation', 5th International Symposium on Robotics in Construction, Tokyo, Vol 2, pp703-708

5 Matheus, C.J., 1986, 'The Internals of FORPS: A FORTH-based Production System', Journal of Forth Application and Research, Vol 4, No 1, pp7-27

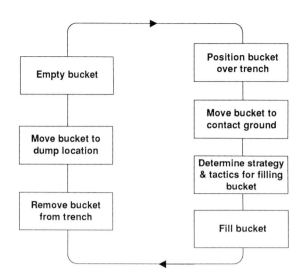

Fig 1   Sequence of operations involved in trenching

Fig 2   JCB 801

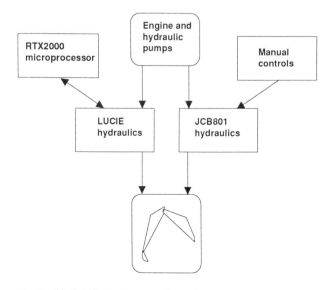

Fig 3   Hydraulic system configuration

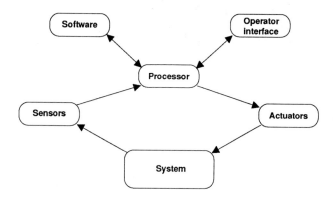

Fig 4 The general configuration of a mechatronic system

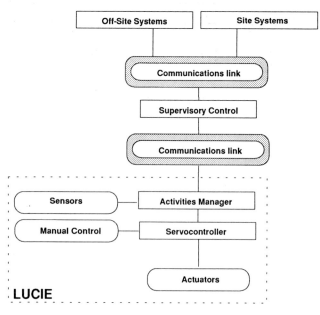

Fig 7 Control system hierarchy

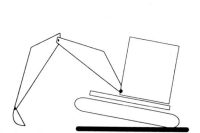

Fig 5 Tilting of excavator cab as a result of reaction of bucket on ground

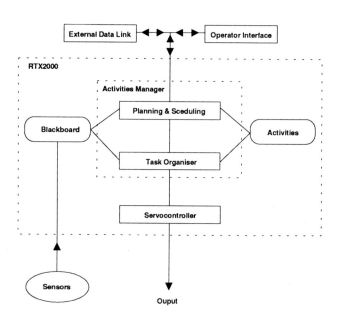

Fig 8 LUCIE software modules

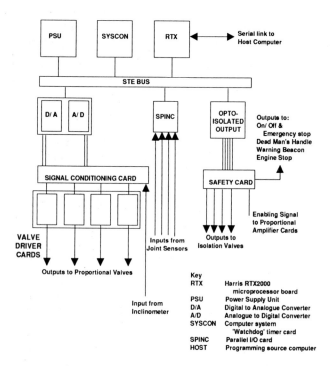

Fig 6 LUCIE electronic system

Experience of mechatronics applied to the design and management of oil-hydraulic and pneumatic systems

S FERRARIS, A LUCIFREDI, E RAVINA
Universitá di Genova, Italy

SYNOPSIS The paper refers on experiences of mechatronic approaches mainly oriented to functional design, management and control of oleotronic and pneumotronic systems. Some aspects of a wide research activity to investigate on the synergetic combination of precision mechanical components, electronic control and computing in fluid power systems are described.

1 INTRODUCTION

The integration between mechanical and electronic components becomes more and more usual also in fluid power applications, improving the characteristics of the overall system.
The mechatronic approach in the fluid power field involves themes of mechanical engineering (actuators design and manufacturing, system dynamics,...), control (control system design, real time systems,...), computer science (algorithm implementation, artificial intelligence applications,...), electronics (sensors, hydrid devices,...). A recent inquiry of U.S.A. National Fluid Power Association concerning issues related to the fluid power industry [1] has pointed out that there is a unanimous feeling that in the next years ther will be an encrease in the mechatronic applications activities, particularly relevant in system integration, CAD approaches, diagnostics and sensoring. Moreover, from the educational point of view significant improvements are needed to make engineers and designers aware of the industrial significance of the mechatronic approach.
At the engineering faculty of Genoa the authors are developing a wide multidisciplinary research and didactic activity on mechatronic applications in the fluid power field, investigating on the synergetic combination of precision mechanical components, electrical devices, electronic control and computing, showing the implicance of a integrating discipline as mechatronics within the design, control and management of oil-hydraulic and pneumatic systems.
Experiences on oleotronic and pneumotronic components and systems involve theoretical and experimental aspects. An original modular package operating on personal computer and conceived to solve problems related to modelling, simulation and identification of components and systems has been implemented. Approaches of control and management of advanced oil-hydraulic and pneumatic systems interfaced with programmable logic controllers and/or with dedicated electronic cards have been investigated.

2 MODELLING AND SIMULATION

The analysis and the optimization of a fluid power integrated design require specific implementations of non-standard modelling and simulation procedures. Very different aspects of interest are involved: in particular

i) the functional description of the circuit and its test of conformity to the required operating sequences;
ii) the sizing and choice of hydraulic and pneumatic devices and of other circuital elements;
iii) the dynamic simulation of single components and/or subsystems;
iv) the comparison among different circuital solutions and between traditional and "smart" components relating to the definition of the optimal configuration;
v) the optimal selection;
vi) the problems related to the maintenance.

All these features can be analyzed through an original integrated package developed by the authors on personal computer. The various options can be performed separately or in integrated way following totally free sequences defined on-line by the user.
By means of the circuit description option very different typologies of circuits (oil-hydraulic, pneumatic, electro-hydraulic, electro-pneumatic, oleotronic, pneumotronic) can be studied. Within a general purpose CAD environment customized by the authors for fluid power applications the user has available various parallel menus of symbols corresponding to devices of different nature (mechanical, electrical, electronic). The menus structure is totally open and new symbols can be generated in addition to the existing ones. Following an interac-

tive procedure supported by editing options the functional scheme can be described starting from the description of the sequential chart related to the automatic sequence of interest. Fig 1 shows an example of circuit description option concerning a circuital solution of a position automatic control using an electro-hydraulic servocylinder.

On the same CAD graphic page the block regulation scheme and the corresponding circuit have been generated; in this case the regulation scheme includes programmer (1), transducer (2), amplifier (3), servovalve (4), power supply (5), oil-hydraulic cylinder (6), load (7). In the corresponding oil-hydraulic circuit the servocylinder materializes the blocks 4 and 6.

The availability of different menus allows also the circuit generation of dedicated control electronic cards and the representation of "ladder" schemes when electro-hydraulic or electro-pneumatic systems must be controlled by PLC. Original interface modules make possible the recalling of these schemes during the PLC programming phase.

Fig 2 shows the CAD scheme of an electro-pneumatic circuit, simulating an automatized pneumatic workstation, and the corresponding ladder diagram. All the circuits described within this environment can be animated in order to verify their correctness before organizing the real system.

Pointing on screen elements or subsystems inserted in the functional scheme, sizing and/or simulation procedures can de automatically selected. Within the sizing option the pointed component is automatically recognized and the corresponding sizing code is reached within a modular library of programs and started. Fig 3 reports a sizing phase of the servovalve inserted in the scheme of Fig 1. The sizing option also manages the selection procedures of components available on the market, allowing the access to an archive of informatic catalogs. That is particularly useful for instance during the selection of sensors and transducers because the comparison of the different characteristics is strongly simplified.

Besides, an option concerning simulation is available. Starting from the automatic recognition of the component or of the subsystem selected on the screen, a procedure transparent to the user activates a dynamic simulation environment managing an archive of original mathematical models corresponding both to single components and to preassembled solutions.

Different simulation approaches can be developed: in particular space states analysis and bondgraphs techniques have been implemented [3,4].

Fig 4 synthesizes some results concerning the dynamic simulation of a jet pipe servovalve in different operating conditions, varying the exciting current : step exitation (Fig 4a), pulse (Fig 4b), ramp with different slopes (Fig 4c and 4d).

The subsystem simulation is generated by automatic assembling of the mathematical models corresponding to single elements: the dialogue between CAD and simulation environments is particularly useful to test a very large number of circuital solutions, changing elements and/or forcing functions. In fact the user can easily substitute a component with another one within the CAD scheme and restart with the simulation : the comparison of the dynamic responses guides to the optimal selection. Besides, with particular reference to the automatic fluid power circuit design, the possibility to test a large number of circuital configurations allows another optimization aspect concerning the reduction of the number of components necessary to perform the automatic sequence.

Finally, within the integrated environment proposed by the authors, problems related to maintenance and optimal selection of devices can be analyzed using artificial intelligence software supports. An hyperstructure oriented to selection and maintenance of mechanical components have been implemented [5]. Since the selection and maintenance procedures often follow non standard sequences, the hypertexts approach makes possible an user-friendly management of this type of problems, allowing the organization of texts, images, notes, graphs, etc. in a very flexible way. The hyperstructure also manages an external code based on neural networks, customized by the authors for selection and maintenance procedures, allowing the solution of problems based on conflictual information.

3 SOME MECHATRONIC APPLICATIONS

The integrated package described above is interfaced with management and control software of specific oleotronic and pneumotronic systems. The information deduced by data acquisition procedures applied to real circuits are stored on disk and can be used during the simulation phase in order to update characteristic parameters and/or to validate the theoretical models. In addition, during the PLC's programming phase the user can recall in parallel the "ladder" representation generated within the CAD environment, making easier the programming procedure.

Two example of mechatronic applications are breafly described hereafter.

Fig 5 illustrates an automatic system simulating a pneumatic pick-and-place workstation controlled by a SQUARED SY/MAX 400 PLC (Fig 6), with digital and analog I/O, high speed control modules and network module. In addition to the local control the authors have implemented the remote control via modem. The local station (laboratory of Meccanica generale e meccanica delle vibrazioni of the Institute of Meccanica applicata alle macchine, Genoa) is connected via modem with the remote station (workshop of SQUARE D Company in Arenzano, 20 km far from Genoa). The remote station controls automatically the pneumatic system by means of an interfacing code for synoptic representation, showing the current state of the circuit through animated sketchs and allowing also the development, by a remote user, of tests on working and emergency conditions.

The second application example concerns a "smart" oil-hydraulic workbench (by DUPLOMATIC, Fig 7). Specific aspects of this research activity involve the assessment of the actuators performances in terms of positioning accuracy, repeatability errors, maximum velocity and acceleration, etc. The system essentially consists on two oil-hydraulic linear actuators having LVDT position transducers, driven by proportional valves interfaced with dedicated electronic cards for the local control and with PLC for the remote control. The actuators can operate separately or in "master / slave" configuration. Fig 8 shows a phase of the remote control where the use of two personal computers allows the parallel management of the ladder sequence and the synoptic representation.

The research activity on pneumotronics and oleotronics involves many other precision mechanical components and microprocessor based electronic devices. Different types of PLC (SQUARE D, Telemecanique, FESTO,...) are available and different programming language (boolean, evolute ladders, GRAFCET, FUP, BASIC, C,...) can be compared by the user with reference to the characteristics of the automatic sequences of interest. Other specific activities under development concern "smart" pneumatic linear actuators driven by innovative on-off valves set and controlled by dedicated electronic card and an advanced precision workstation based on rodless pneumatic cylinders. The detailed description of these activities and the corresponding results will be presented in a next paper.

4 CONCLUSIONS

The previous considerations have shortly presented some experiences of the authors in the field of oleotronics and pneumotronics. The multidisciplinary approach has pointed out the advantages of the mechatronic concept application in the fluid power system design and management. The integrated package proposed seems to be also an useful tool to show mechatronics in education. The interfacing with real systems allows the development of different and original experimental tests.

REFERENCES

(1) NFPA Reporter "NFPA trend survey", nov/dec 1990 edition.

(2) WATTON J. "Fluid power control systems: some mechatronics issues", FLUICOME 91, S.Francisco, U.S.A., 1991

(3) FERRARIS S:, LIN X., LUCIFREDI A., RAVINA E. "Advanced CAD procedures for modelling and simulation in the oil-hydraulic and pneumatic field", IASTED Intl. Symp. MIC 90, Innsbruck, Austria, 1990.

(4) FERRARIS S., LUCIFREDI A., RAVINA E. "Tecniche innovative nella progettazione meccanica", Automazione Integrata, n.11, 1991.

(5) FERRARIS S., LUCIFREDI A., RAVINA E. "Advanced procedures oriented to the selection and maintenance of mechanical components and systems", 9th. Intl. Sypm. AI, Innsbruck, Austria, 1991.

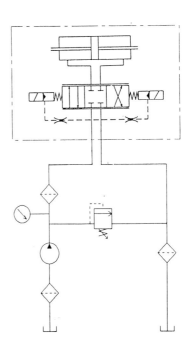

Fig 1 Example of circuit description option

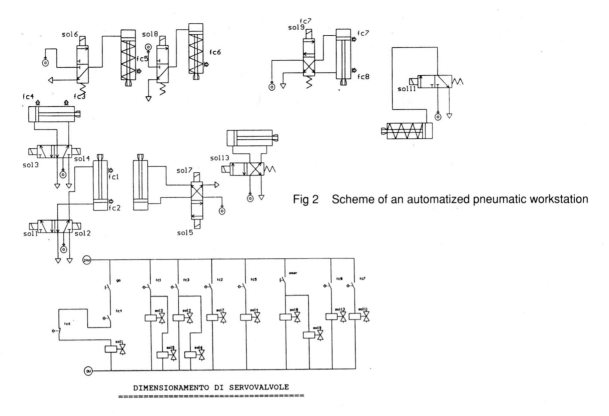

Fig 2   Scheme of an automatized pneumatic workstation

```
DIMENSIONAMENTO DI SERVOVALVOLE
===============================

****     PARAMETRI DI INGRESSO     ****

CARATTERISTICHE DEL CARICO :

Tipo di attuatore                    [Lineare o Angolare] :? L
Forza resistente massima                          [daN] :? 120
Velocita' massima                               [m/min] :? 15
Diametro di alesaggio del cilindro                [mm] :? 50

CARATTERISTICHE DELLA SERVOVALVOLA NECESSARIA ALL'AZIONAMENTO
=============================================================

PORTATA NOMINALE DELLA SERVOVALVOLA     =    40.00    [litri/min]
PORTATA ASSORBITA DALL'UTENZA           =    33.85    [litri/min]
PRESSIONE NOMINALE DELLA SERVOVALVOLA   =    70.00    [bar]
CADUTA DI PRESSIONE NELLA SERVOVALVOLA  =    50.14    [bar]
PRESSIONE DI ESERCIZIO DELLA CENTRALINA =    41.02    [bar]
```

Fig 3   Sizing phase of a servovalve

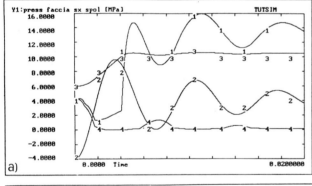
a)

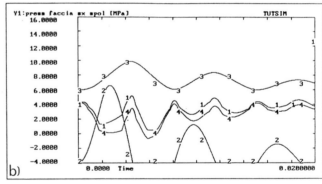
b)

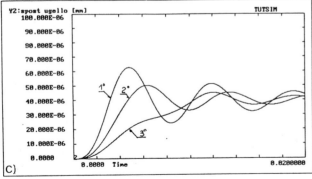
c)

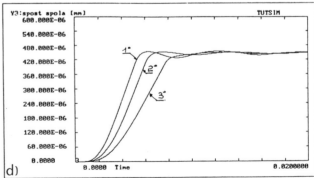
d)

Y1 = left face pressure   [MPa]
Y2 = jet displacement   [mm]
Y3 = spool displacement   [mm]
Y4 = right face pressure   [MPa]

Fig 4   Examples of dynamic simulations

Fig 5　Pick-and-place pneumatic workstation

Fig 7　Oil-hydraulic workbench (DUPLOMATIC)

Fig 6　SY/MAY 400 SQUARE D PLC

Fig 8　Phases of remote control

Machine intelligence: the key capability of mechatronic products

G RZEVSKI, Dipl Ing, MBCS
The Open University, UK

SYNOPSIS

Mechatronics was conceived in the '70s by Japanese as a subject that integrates mechanical and electronic technologies. Since then electronic technology has radically changed and offers now unparalleled opportunities for gaining competitive advantage by improving product performance, making products environment-friendly and simplifying their operation and maintenance. In the early '90s the new trend is to built into machines artificial intelligence modules capable of taking over some of the decision making from their operators. Many examples that underline this trend are discussed, including appliances and leisure equipment recently launched by leading Japanese manufacturing companies Matsushita Electric, Akai and Sharp.

1 INTRODUCTION

There exists convincing evidence of a new trend in the design of machines such as machine-tools, high-speed industrial machines, aircraft, cars, industrial vehicles, appliances and cameras. The trend is to built into machines *advanced information processing capabilities* with a view to improving their performance, make them more environment-friendly and easy to use. In particular, the trend is to make these information processing resources so advanced that they enable machines to operate, at least to a certain degree, *autonomously*, i.e. independently from their operators, and may thus be considered as being *intelligent* (1). In this paper I shall refer to such artefacts as *intelligent machines*.

As an illustration of some aspects of this trend let us look at current developments in the automobile industry. In the late 1980s European car manufacturers launched a joint research programme, Prometheus, with the aim of improving, by the year 2010, car safety by 30% and road traffic flow efficiency by 20%. Within the first two years the following prototypes based on information technology have been developed:

- an enhanced driver vision system based on the use of ultraviolet and infrared light capable of improving visibility in the dark and in fog;
- a sensor-based cruise control capable of maintaining a safe distance between cars;
- a collision avoidance system capable of taking over the control of a car in an emergency;
- a cooperative driving system which enables drivers of cars and heavy trucks to communicate with each other in difficult traffic conditions;
- an automatic route guidance system;
- a comprehensive commercial vehicles fleet management system which included capabilities of determining vehicle positions;
- a head-up display of information based on holograph technology;
- a steer-by-wire car with an aircraft-type joystick instead of the conventional steering wheel;
- a self-parking car.

In addition to Prometheus, there are several other large research programmes in Japan, USA and Europe concerned with developing comprehensive road traffic systems which include land-based and on-board digital hardware and software for traffic information and control, parking reservation and payment and road charging, see references (2) and (3), as well as expert systems offering route selection advice to drivers.

In all new projects independent systems are being replaced by *distributed information systems* which, by means of data communication networks and locally placed microprocessors integrate all control activities within the product. For example, advanced car management systems, currently under development, integrate all previously independent monitoring and control functions such as fuel injection, anti-lock brake control, power-steering, cruise control, monitoring of engine temperature, control of steering, active suspensions, detection of obstacles and control of power transmission by data highways and localised computational resources.

Manufacturing industry provides another interesting example. Until recently production automation relied almost exclusively on rigidly structuring the manufacturing environment, as exemplified by production lines, and using machines programmed to perform repeatedly exactly the same operations. Changes in patterns of demand, which created pressures for shorter concept-to-market lead times and for more frequent improvements of products, coupled with the availability of advanced information processing resources, are beginning now to change the balance. The cost of rigidly structuring manufacturing environments is now prohibitively high in comparison with the cost of developing flexible manufacturing systems populated by intelligent machines capable of coping with unpredictable events. Capabilities of intelligent machines tools now under development include collision avoidance, self-diagnosing, preventive self-maintenance and rapid self-retooling. Distributed production planning and control systems are likely to replace centralised databases, manufacturing resource planning systems, schedulers and controllers.

Elements of intelligence are now designed into a wide range of domestic appliances and leisure equipment with a view to, specifically, simplifying their use and eliminating the need for complex operating instructions. For example, in 1990 a Japanese company launched a washing machine which is capable of selecting the appropriate washing programme based on reasoning about quantity and quality of dirt, size of load and type of detergent used, for a price not much above that asked for a similar conventional appliance. At the same time, the same company developed an intelligent vacuum cleaner capable of adjusting the suction power to suit the prevailing room conditions. A year later, another Japanese company built intelligence into video recorders to achieve a uniformly high quality of images over a wide range of different tapes and yet another launched a camcorder capable of learning from its user how to adjust the aperture to avoid underexposure of objects illuminated from behind.

A thorough analysis of these and many other examples shows that, as a rule, designers of mechatronics products tend to replace, whenever possible, mechanical technology by electrical, electronic, software and artificial intelligence technologies and to expand the use of these new technologies with a view to taking over some of the functions from the operator. This is a gradual process characterised by the following more or less clearly defined stages.

1. Mechanical power transmission lines are replaced by electrical ones, as in manufacturing workshops and diesel-electric locomotives.
2. Machines are designed to take over from the operators some critical monitoring and control functions by building into them digital information processing capabilities and, in many cases, sensors. These machines are capable of following precisely any given computer program, as in conventional robots and computer-controlled machine tools, or maintaining given behaviour under conditions of variable external factors using closed-loop control technology, as in industrial furnaces, new generation high speed packaging machines and aircraft.
3. Mechanical information transmission linkages between the operator and executive parts of the machine are removed and new ones built using digital technology, as in fly-by-wire aircraft and steer-by-wire cars.
4. Further information processing resources and more powerful sensors are designed into machines enabling them to make decisions under conditions of considerable uncertainty and thus exhibit rudimentary intelligent behaviour, as in self-parking cars and collision avoiding mobile robots.
5. Artificial intelligence techniques are used to enable machines to be more or less independent from the operator, as in appliances and leisure equipment described at the beginning of this paper. More advanced intelligent machines are designed to continuously improve their own performance by learning from experience.
6. Machines with information processing capabilities are interconnected into systems with a view to offering new services to customers, as in computer integrated manufacturing systems, intelligent building systems and new proposed road traffic systems.

Two factors will ensure the key role of machine intelligence in mechatronics for the next decade: first, the increasing demand for high value added products capable of meeting rapidly changing requirements and expectations of customers and more stringent environmental specifications and second, the continuously falling price/performance ratios of digital technology in contrast to rising costs of material and material processing technologies.

2 MACHINE INTELLIGENCE

In this section I shall define closer what I mean by terms 'intelligent machine' and 'machine intelligence'.

Let us start from the beginning. The work and maintenance of every machine has to be planned and controlled. In other words, machines have to be *managed*. Machine *management* includes setting machine goals and performance levels, scheduling its load within constraints imposed by external factors, planning its maintenance and controlling its behaviour during operation.

The simplest machines, such as present mass-produced cars and appliances, are managed by their operators, i.e. their users. More advanced machines, as robots and computer-controlled machine tools, are *programmed* to behave in a particular manner. The pattern of behaviour of such machines will change only if their control program is replaced.

There are, however, machines with built-in self-management systems that are capable of modifying their own behaviour, at least to a certain degree, whenever there is a need to accommodate an unexpected change. These machines are referred to throughout this paper as *intelligent*. The Open University has formed a research laboratory for investigating methods for designing intelligent machines (4) and a team of academic members of staff is developing distance learning material in the same subject area. For the purposes of our research and teaching we define *intelligence* as *a capability of a system to achieve a goal or sustain desired behaviour under conditions of uncertainty*.

The usual sources of uncertainty are:

- the occurrence of unexpected events, either internal, such as a component failure, or external, such as an unpredictable change in the world in which the system operates;
- incomplete, inconsistent or unreliable information available to the system for the purpose of deciding what to do next.

The intelligence as defined above may be considered as basic. Higher levels of machine intelligence would include an ability to learn from interaction with the world in which the machine operates, an ability to shape its environment with a view to achieving specified goals and, possibly, a capacity for formulating own goals. Intelligent capabilities such as forming new concepts, creating new systems or self-expression are not discussed in this paper.

The word intelligence has emotional connotations and some people resent the term being applied to machines. In this paper the term is used in a very pragmatic manner as a means of distinguishing machines that are capable of making decisions under conditions of uncertainty from machines that are built, or programmed, to carry out repetitive tasks and are capable of changing their behaviour only if instructed by their operators to do so. Such usage of the term intelligent is now generally accepted, although the details of proposed definitions may vary.

Under deterministic, i.e. predictable, conditions it is much more effective to have a programmed rather than an intelligent machine. Major strengths of programmed behaviour are precision and repeatability. The major weakness is in its inability to cope with unexpected events.

Intelligent machines must have a considerable capacity for collecting, storing, processing and distributing information. An intelligent machine receives information from its environment through sensors, such as:

- laser bar-code readers,
- photo-sensitive cells,
- video cameras,
- radars,
- sonars,

and sends information to its environment by means of signalling, broadcasting, printing or display equipment, such as:

- flashing lights,
- sirens,
- infrared devices,
- printers,
- display panels,
- screens.

The acquired information is, as a rule, stored in digital memories, processed by digital processors and transmitted through digital communication networks.

Intelligent machines may operate in a self-contained mode or be linked into systems. Examples of systems of interconnected intelligent machines include:

- computer integrated manufacturing systems, and
- road, rail and air traffic control systems.

The important feature of a system is that its internal links modify performances of its elements so that the overall system performance is better than the sum of performances of elements: a phenomenon called the *emergent property* of a system.

There is a wide variety of application areas open to intelligent machines. Let me list some of them:

- manufacturing,
- tunnelling and underground work,
- civil engineering and construction work,
- transportation by road, rail, sea and air,
- fire-fighting and emergency rescue,
- nuclear engineering and operations,
- underwater operations,
- space missions,
- processing of materials that are hazardous, toxic, pure, expensive or unstable,
- preparation, handling and packaging of food,
- agricultural work,
- collecting and monitoring of environmental data,
- off-shore oil and gas exploitation,
- microtechnology,
- medicine and health care,
- leisure.

3   SELF-REGULATION

The most elementary behaviour that could be arguably classed as intelligent is *self-regulation*. It denotes the capability of a machine to achieve and sustain the desired behaviour when working in an environment which changes in time in a limited way. The characteristics that change (called variables), the range of measurable changes, and the way in which the machine should respond to any particular change are known in advance. Only the timing and magnitudes of changes are not known. Essential components for self-regulation are:

- sensors that collect information about the actual state of the machine and its environment;
- a decision mechanism which selects an action required to achieve or maintain the desired behaviour;
- effectors that move the machine to the desired state.

In general, for the purposes of self-regulations a machine may monitor one or several of measurable physical characteristics, called variables, such as:

- position,
- distance from a given object,
- direction of movement,
- speed,
- acceleration,
- pressure,
- liquid level,
- thickness,
- composition.

Whatever the variable or the set of variables, the mechanism of self-regulation is always the same: the *feedback loop*:

1. *Measure* values of variables selected to represent a machine behaviour by means of sensors.
2. *Compare* the measured values with the desired values.
3. *Decide* which action to take in order to eliminate the observed difference, if any.
4. *Implement* the selected action.
5. Go to step one and *obtain feedback* on results of the action.

Auto-focusing of a camera and auto-piloting of an aircraft are examples of self-regulation.

4 REASONING AND KNOWLEDGE

To cope with an *unstructured* environment, where variable characteristics are not measurable (e.g. the difference between a person and a piece of furniture in a room in which a mobile robot must operate), where several characteristics change simultaneously and in unexpected ways, and where it is not possible to decide in advance how the machine should respond to every combination of events, a higher level of intelligent behaviour than self-regulation is required. The machine needs a capability to reason. Reasoning can only be done if the machine has access to *knowledge*, e.g.:

- knowledge about its goals and tasks;
- knowledge about its own capabilities;
- knowledge about the environment in which it operates.

Then, *reasoning* denotes a search through the knowledge space for that element of knowledge (e.g. a rule) which helps the machine to decide what

action it should take to compensate for, or take advantage of, the change. Reasoning is also required for planning future behaviours when knowledge about the future is incomplete or unreliable.

Some elements of knowledge required for generating intelligent behaviour have to be collected by machine designers from various sources, including the expertise of human operators who have performed such or similar tasks successfully, and written into a, so called, *knowledge base* using an appropriate computer-readable language. The rest must be acquired by the machine using its sensors and information processing resources.

To illustrate reasoning let us consider an example of an autonomous vehicle transporting goods along corridors of a store.

The vehicle is likely to have a model of its own sensors and effectors and a map of the world in which it operates, which may include knowledge about all available corridors, their lengths, widths and permissible speeds, and even about kinds of objects that may represent obstacles for the vehicle. As it travels, the vehicle observes its world, perhaps by means of video cameras, sonars and laser rangers, and compares what it 'sees' with what is on its internal map. It *navigates* through passages and among obstacles. If it detects an unexpected object within its path (e.g. a person or another vehicle), it may attempt to identify the object by comparison of its observed characteristics with characteristics of objects that it knows, and it may attempt to determine whether the object is moving and if so then in which direction, with what speed, etc. If collected information is incomplete, the machine will make an informed guess about the detected object and will engage in planning an appropriate sequence of action, such as sounding a warning, slowing down, or changing direction. During the planning activity a suitable action is selected by searching through machine knowledge base for all available alternatives and, if the situation allows, the preference is given to actions that minimise the disruption of the schedule. Finally, the selected action will be implemented and a feedback message on the result of the selected action will be obtained, thus closing the control loop.

The similarity with self-regulation is obvious. There is a closed loop which includes monitoring, comparison with a standard, decision making and implementation, followed by a feedback. There are, however, several major differences:

- raw sensory data received from a variety of sensors must be analysed and interpreted;
- the internal model of the world must be continuously updated;
- decisions on the next action must be made under conditions of uncertainty caused by the occurrence of unpredictable events or by incomplete, inconsistent or unreliable information;
- future actions must be planned with a view to minimising the disruption of the schedule;
- often, communication with other machines operating in the vicinity is required.

Methods for making decisions under conditions of uncertainty are based on the information processing theory of mind developed by Newell and Simon (5) or on Zadeh's fuzzy logic (6).

5 LEARNING AND PATTERN RECOGNITION

Learning behaviour may be created by making use of artificial neural networks. An *artificial neuron* is an information processing unit which models the behaviour of of brain cells called neurons. A human brain has approximately 10 to the power of 10 neurons each connected to a large number of other neurons. Biological neural networks process, in parallel, electrical signals generated by chemical processes. They store information by modifying strengths of neuron connections, which means that information is stored in a distributed fashion. Neural networks are capable of learning patterns of information and recognising them even if incompletely specified. Thus artificial neural networks may be used to provide machines with capabilities to learn as they operate and to generate intelligent behaviours through the process of pattern recognition. This eliminates or at least reduces the task of collecting knowledge relevant to machine operation and entering it into the knowledge base.

An interesting example of a machine capable of learning is a state-of-the-art camcorder. When a strong light shines behind a person and, as a result, his or her face is in the dark, the conventional camcorders (those equipped with self-regulation) cannot obtain a good quality image automatically. The user must intervene and adjust the appropriate control manually. In contrast, a neural network camcorder will learn, over a period of time, how to adjust controls automatically by associating patterns of lighting and patterns of manual adjustments carried out by the user. Thus, the camcorder learns how to cope with various filming conditions by monitoring the behaviour of its own user.

Fundamentals of neural networks are covered, for example in (7).

6 AUTONOMOUS INTELLIGENT AGENTS

Intelligent behaviours of a machine may be created through an interaction of *autonomous intelligent agents* resident in the machine, each having its own independent goal. The overall behaviour of the machine is then a result of a synergy of behaviours of constituent agents.

For example, a machine tool may include two independent intelligent agents, one charged with a goal of achieving the fastest practical speed of cutting a work-piece and the other with a goal of maintaining the machine in the best possible working order. Under conditions of normal operation, the first agent will monitor speed of cutting and maintain it at the highest acceptable level whilst the second agent will be monitoring the tool wear. When the tool wear reaches certain limit the second agent may decide that unless the cutting speed is reduced the tool will break in the middle of the current operation and will send a request for slow-down. Conflicting demands generated by these two agents may be resolved by negotiation (which computers, in contrast to people, can accomplished in a fraction of a second) or by an independent arbitration mechanism.

Autonomous intelligent agents are a very active area of research in artificial intelligence and is likely to remain so in the foreseeable future. The whole idea of replacing central control by a group of autonomous actors that negotiate a consensus is eloquently presented by Minski in his book The Society of Mind (8).

7 ARCHITECTURE OF INTELLIGENT MACHINES

Since the emergent properties of systems are created by linking system components in particular ways, a whole subject of study of system interfaces has recently sprung into existence, termed the study of system architectures.

The *architecture* of a system is a set of rules that define interfaces between the system and its environment and between system elements. For example, the architecture of a building defines the way in which windows and the roof interface with elevations and the way in which the whole building fits into its urban environment. Similarly, the architecture of a computer defines the ways in which hardware is interfaced with the operating system.

Our concern here is with architectures of intelligent machines and, in particular, the dependence of the overall machine intelligence on the way in which its elements are interconnected.

An intelligent machine has to carry out a complex set of tasks and therefore requires an effective architecture. Several possible architectures have been tried by mechatronics designers, including hierarchies, networks and layers.

In *networks,* in contrast to hierarchies, there are no levels of importance and all element may be connected to each other, as in an 'old-boy networks' or, alternatively, into rings or stars, as in communication networks. Network architectures are used when there is a need for cooperation between units that are equal in importance but different in terms of skills or capabilities, as in distributed computer systems, or if there is a requirement for parallel processing of information, as in neural networks .

Layered architectures consist of self-contained elements, called layers, each connected to a set of inputs and outputs and thus each capable of creating a system behaviour (9). Since behaviours generated by different layers may be in conflict with each other, as in the example given in Section 1.5, the layered architecture may include an arbitration mechanism responsible for ensuring a proper overall system behaviour. An important characteristics of layered architectures is that they are ideally suited for the incremental development of systems.

A network architecture which includes perception, cognition, execution, self-maintenance, and energy conversion subsystems is selected by the Open University mechatronics team as the most appropriate for discussing various aspects of intelligent machines. This architecture is shown in Figure 1.

The role of the *perception* subsystem is to collect, store, process and distribute information about the current state of the machine and its environment. To do that effectively the perception subsystem may be required to construct, store and update models of the machine and its environment.

The *cognition* subsystem is responsible for evaluating information collected and processed by perception and for planning actions that will take place in the near or distant future. The implementation of the cognitive function is at present computationally expensive and time consuming and therefore many types of intelligent machine are designed to operate without the capability for planning their own behaviour.

The *execution* subsystem is responsible for controlling all activities of a machine, based on instructions normally received from cognition (purposeful behaviour) or, in situations when there is a need for a rapid avoidance action, directly from perception (reactive behaviour). The flows of information underlying the purposeful and reactive behaviours are depicted in Figure 2. In current intelligent machines, as a rule, reactive behaviour is predominant and cognition plays a relatively small role in comparison with perception and execution.

The *self-maintenance* subsystem is responsible for maintaining the machine in good order during its normal operation, which includes intermittent monitoring of the behaviour of key parts with a view to discovering a fault before it occurs (preventive self-maintenance), or immediately after it occurs (self-diagnosing). It may also carry out self-repairs by means of reconfiguration aimed at removing the faulty part, and in very advanced machines, it may have a responsibility for continuously improving machine performance. The following are examples of behaviours which are considered as self-maintenance:

- when a car management system monitors wear of brake pads and, based on this information, makes decisions how to compensate for wear and when to replace the pads (preventive self-maintenance);
- when a printer signals the presence of a fault and identifies it (self-diagnosis);
- when a spacecraft re-configures one of its control systems in order to eliminate a faulty component (self-repair);
- when an autonomous transportation vehicle learns from experience which routes through a factory are least congested at a particular time of the day and selects its trajectory accordingly (self-improvement).

To carry out its main function a machine must have an appropriate energy supply and, to maintain itself in good order, it must eliminate undesirable waste, such as heat. These tasks are carried out by the *energy conversion* subsystem whose function is self-explanatory and would not be discussed any further.

It is important to understand that all subsystems described in this section are conceptual rather than physical. In other words, although it may be useful, in abstract, to distinguish the task of processing sensory information from that of controlling physical movement of a machine, this does not imply that these two tasks cannot be done by the same computer. Conversely, although it is quite effective to mount tactile sensors on robot grippers, this in no way implies that functions of these two devices are conceptually the same.

8 CONCLUSIONS

Information technology offers new opportunities for improving performance of machines and making them more environment-friendly and easy to use. The most interesting opportunities may be realised by building into machines some form of artificial intelligence which enables them to achieve given goals and sustain desired behaviour even if the world in which they operate is subject to unpredictable (within certain limits) changes.

REFERENCES

(1) RZEVSKI, G., Strategic Importance of Engineering Design. <u>Journal of Design and Manufacturing,</u> 2, No 1, 1992.

(2) ARAI, H., New Roles of Automotive Electronics. Proceedings Vol 1, ISATA 22nd Int Symp on Automotive Technology & Automation, Florence, 1990.

(3) SODEIKAT, H., How can the Dynamic Route Guidance Systems ALI-SCOUT Help to Solve Traffic Problems and to Promote the Enforcement of Public Traffic Policy Measures? Proceedings Vol 1, ISATA 22nd Int Symp on Automotive Technology & Automation, Florence, 1990.

(4) RZEVSKI G., On the Design of Intelligent Machines: Critical Issues and Approaches. Proceedings of Int Conf on Engineering Design, Hawaii, 1991.

(5) NEWELL, A. AND SIMON, H. A., <u>Human Problem Solving</u>. Englewood Cliffs, Prentice-Hall, 1972.

(6) ZADEH, L. A., A Theory of Approximate Reasoning. <u>Machine Intelligence,</u> 9, 149-194, 1979.

(7) ALEXANDER I., MORTON, H., <u>An Introduction to Neural Computing</u>. Capman & Hall, 1990.

(8) MINSKI, M., <u>The Society of Mind</u>. Heinemann, 1985.

(9) MAES, P. (ed), <u>Designing Autonomous Agents</u>. MIT/Elsevier, 1990.

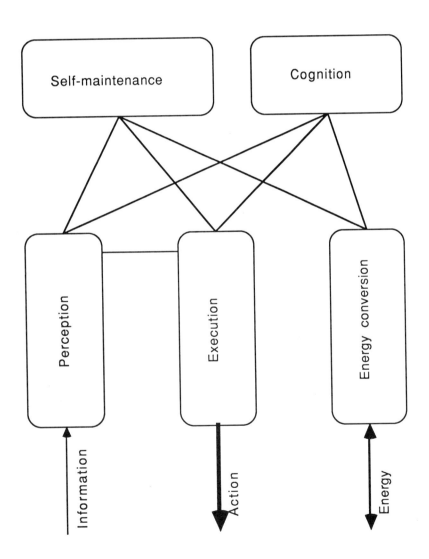

Fig 1

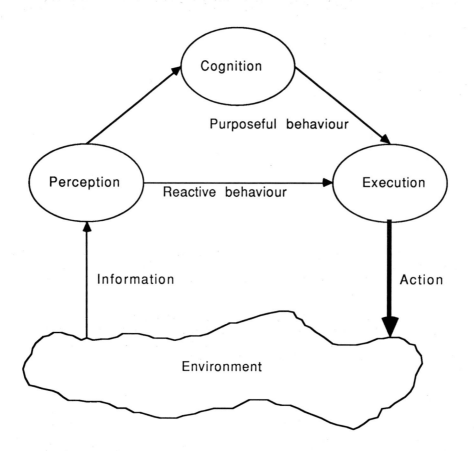

Fig 2

Mechatronics – meeting the technician's needs into the 21st Century

C C B DAY, BSc, MSc, PhD, CEng, MIMechE
Gwent College of Higher Education, UK

SYNOPSIS

Over the past decade, the UK has experienced a loss in its manufacturing ability. To ensure a manufacturing base into the 21st century there are a number of changes that need to be accomplished by educationalists and industrialists.

This paper discusses the model used by the Business and Technology Education Council (BTEC) in developing a Higher National Mechatronics programme for engineering technicians and how it is perceived by industrialists to meet their present and future needs.

1.0 INTRODUCTION

The output from the manufacturing industries of the United Kingdom have, over recent years, considerably reduced. Factors that have contributed to the erosion of our manufacturing base are lack of investment in both equipment and training, together with poor national economic performance. These factors have resulted in a consequent inability to attract young people into the engineering profession. The image, which is certainly confirmed by recent government statistics (1), is one of company closures and a continual decline in the number of employees in the industry, fig. 1

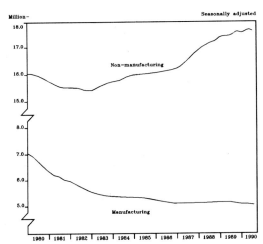

Fig 1 Manufacturing and non manufacturing employees in employment
(Reproduced from *Employment Gazette*, April 1991)

Further evidence of the decline is seen in a fall in output and an increase in the cost per unit of output, fig. 2. All of these factors point towards a worrying reduction in our competitiveness in world markets.

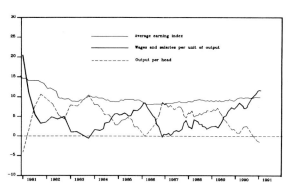

Fig 2 Earnings and output per head : manufacturing industries – increases over previous year (Reproduced from *Employment Gazette*, April 1991)

This reduction in our "competitive edge", particularly when compared to the economic performance of the Japanese, does give cause for concern. They are seen to be able to rapidly transform new ideas into products. This ability clearly results from their philosophy of using integrated teams of personnel to provide a multi disciplined approach to product design, manufacture, purchasing, sales and maintenance. This philosophy of producing optimized designs and manufacturing processes linked strongly with R&D at both the customer and supplier ends of the product chain is known as "*Mechatronics*". To try and adopt this philosophy of approach in the UK manufacturing industry would certainly have far reaching implications for Higher Education in the way it presently educates and trains at both vocational and professional levels.

2. EDUCATION AND TRAINING ISSUES

The Department of Trade and Industry have accepted for some time that in order to move towards economic recovery it is essential for industry to establish a management and manufacturing environment which is less traditional than at present and based upon the new technologies. To support this view they initiated a development programme with the Business Technology and Education Council (BTEC) and a Consortium of Northern Colleges to produce teaching modules in Mechatronics.

It was their intention that they would embody the latest technology and management practices. The modules were to be made available to Educational establishments for updating existing teaching programmes and would also be available for in-company updating/retraining activities. Arising out of the early work of the project nineteen modules covering the field of Advanced Manufacturing Technology (AMT) were developed in conjunction with a number of major companies. These modules represented a study time of 570 hours and covered the following subject areas:

- Computer Aided Design and Manufacture (CAD/CAM)
- Programmable Logic Controllers (PLC)
- Electro hydraulic and pneumatic Systems and Devices
- Computer Aided Quality Assurance
- Computer Aided Inspection
- Computerized Numerical Control
- Robot Technology
- Interfacing Systems and Devices

From early discussions between the educationalists and the companies (2) it was evident that for the industrialists needs to be met the following competences needed to be designed into the proposed Mechatronics programme:

- technical skills in integrating across mechanical, electronic, manufacturing, maintenance and software engineering.
- applying knowledge and experience in manufacturing/maintenance planning, control software and software systems
- application of the principles of finance, costing and quality control to the manufacturing process

This was indeed a wide programme specification by any standards but one that fitted well within the Mechatronics definition.

The present UK Higher Education provision at either vocational or professional level would have great difficulty in meeting the demands made within this specification. The majority of educational provision is still very much single subject orientated. There are, however, a small number of Universities at which Mechatronics can be studied at either undergraduate or post graduate levels (3). At the present time there are no vocational programmes in Mechatronics leading to Higher National qualification. It is expected, however, that this situation will be remedied by September 1992.

All BTEC programmes have inbuilt into their curriculum the development of the following common skills.

- Managing and developing self
- Working with and relating to others
- Communicating
- Managing tasks and solving problems
- Applying numeracy
- Applying technology
- Applying design and creativity

Each student will have within their final qualification a profile of their common skills achievements throughout their programme of study. These skills will enhance their capability to manage change, a requirement that will be essential for the future engineer or technician.

The provision of BTEC Mechatronics programmes at Colleges will give technicians who have completed single subject qualifications the opportunity to broaden their knowledge across the traditional academic boundaries. Many technicians are already involved within their companies in working at the interfaces of the technologies. It would therefore be reasonable to assume that they would wish to have their experience recognised in a formal way which would manifest itself in a demand for a Higher National qualification in Mechatronics. Fig. 3 shows the number of BTEC registrations on all engineering programmes since 1984/85 (4).

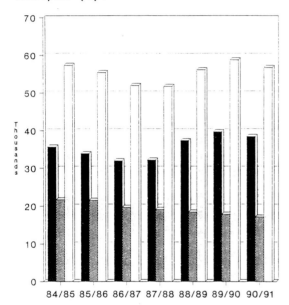

Fig 3 Registrations on BTEC engineering courses (Data derived from 'State of Play Report' Internal BTEC Report)

The diagram reflects the current state of the industry in that registrations on both the National and Higher National programmes have fallen over recent years. However it can be seen that over the past three years there have been an average of 55,000 students registered per year.

Fig. 4 provides details of the number of completed awards at National and Higher National levels. There is no direct correlation between fig. 3 and 4, each figure provides discrete data. Because of the flexibility of access to BTEC programmes cohort analysis is not always possible.

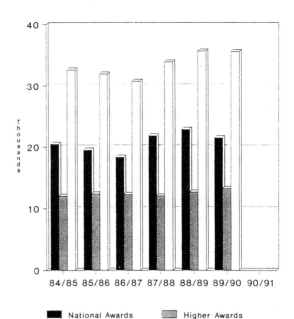

Fig 4 Awards on BTEC engineering courses (Data derived from 'State of Play Report' Internal BTEC Report)

Both figures serve to indicate the possible size of the student market that may be available for either retraining/updating in Mechatronics or studying the programme as a continuation from a National qualification.

To ensure good access possibilities for the programme it will need to be delivered in a flexible manner. The majority of technicians who would be attracted to this vocational programme would already be in employment and therefore mixed mode delivery would be desirable.

3.0 THE BTEC MECHATRONICS MODEL

The model was devised to ensure that the outcomes from the study programme would satisfy the perceived industrial needs articulated in the early joint meetings between educational representatives and the companies involved in the project. These have been distilled into three main overall aims for the programme (5):

- to provide a mix of skills in electronics, computing technology and mechanical engineering

- to produce technicians able to apply multi-disciplined and integrated approach to the design, manufacture or maintenance of products and processes.

- to produce technicians capable of operating and communicating across engineering disciplines.

The programme is based around a core which presents the essential elements of Mechatronics derived from the three major subject areas of computing, mechanical, and electronic engineering. The study of the core material requires 240 hours for a part time mode of study and 360 hours for a full time Diploma programme. The core consists of 4 BTEC modules requiring 60/90 hours of study time. The technical elements embraced within the core modules are

- Transducers and Sensors
- Mechanical, Hydraulic and Pneumatic Actuation Systems
- Electrical Actuation Systems Control
- Interactive Systems
- Diagnostics
- Microprocessor Systems
- Programmable Logic Controllers
- Communication Systems

The Mechatronics core material supports the continuation of the programme into industrial applications of design (more accurately described at technician level as redesign), manufacture and maintenance. The model is represented in Fig. 5.

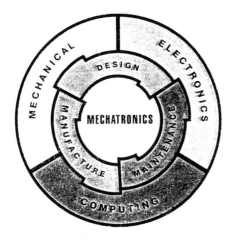

Fig 5 Diagrammatic representation of BTEC mechatronics model
(Reproduced with the permission of BTEC)

The _three_ specific programmes for Mechatronic Applications in Design, Manufacture or Maintenance Engineering allows an educational establishment to reflect the needs of its local companies and its own specialist strengths in the choice of modules on offer. A general Mechatronics programme could also be chosen which would consist of the core modules together with a choice of modules across the three options. To achieve an award a successful performance would need to have been achieved in 10 modules together with an acceptable common skills profile.

The Advanced Manufacturing Technology modules that were developed earlier in the project could now be used to advantage within the option paths of the programme. Fig. 6 indicates typical modules which could be included in the options together with the core modules (6).

Important elements in all BTEC vocational programmes of study are the teaching and learning strategies adopted together with the assessment methods used.

It is extremely important that the teaching and learning strategies adopted for delivery are compatible with the expected outcomes from the programme. To achieve integration of the curriculum content it is expected that the programme will be delivered through multi-discipline assignments which are either college devised or company based but must reflect, strongly, industrial applications. This will require the availability of a facility in which practical work can be undertaken to underpin the integrative nature of the Mechatronics programme.

Gwent College is a partner in the Eurotechnet II programme for the development of the "Factory of the Future". Fig. 7 shows the Mechatronic teaching laboratory that has been developed for either single element teaching such as CNC, DNC, robotics and interfacing or for the availability of a full flexible manufacturing cell. A teaching/development area such as this providing "hands on" experience of the integration of systems should be an essential requirement to ensure quality of the learning experience for the student studying any Mechatronics programme.

Core	
Mechatronics A	2.0
Mechatronics B	2.0
Industrial Studies	1.0
Project	1.0
Common Skills	0.0

Application areas/optional units					
Design		**Manufacture**		**Maintenance**	
Mechanical Science	1.0	Production Planning & Control	1.0	Maintenance of Manufacturing Systems	1.0
Engineering Design	1.0	Quality Assurance	1.0	Maintenance of Material Handling Systems	1.0
Computer Aided Design	1.0	Manufacturing Technology	1.0	Maintenance of Factory Services	1.0
Engineering Materials	1.0	Computer Aided Manufacturing Systems	1.0	Plant instrumentation & Control	0.5
Computer Aided Engineering	1.0	Computer Aided Engineering	1.0	Maintenance Planning	0.5
CADCAM A	0.5	Computer Aided Machining A	0.5		
CADCAM B	0.5	Computer Aided Machining B	0.5		
		Production & Inventory Management	0.5		
Programmable Logic Controllers	1.0	Computer Aided Quality Assurance	0.5		
Electronic CAD	1.0	CADCAM A	0.5		
Digital Circuit Design	1.0	CADCAM B	0.5		
Analogue Circuit Design	1.0				
Computer Aided Engineering (Elecronic)	1.0				

1 MODULE – 60 HOURS LEARNING SUPPORT TIME (PART TIME)
90 HOURS LEARNING SUPPORT TIME (FULL TIME)

Fig 6   Suggested structure for higher national certificate programme

4.0 INDUSTRIAL AWARENESS

The Higher National Certificate programme in Mechatronics was officially launched by BTEC in May 1991 with the publication of a booklet providing guidance on programme delivery and its module content (6).

An awareness Seminar programme was organised at seven educational centres throughout England and Wales to which representatives from Colleges and industry were invited. The intention of the meetings was to expose the Mechatronics programme to a much wider audience than the group of Companies who were involved with its early development. The attendance at the Seminars were disappointing particularly from the invited industrial representatives. The reactions of the participants to the proposed HNC was mixed. Colleges generally welcomed the development whilst the industrialists, reflecting the present economic climate expressed the difficulty in undertaking any investment in training. The resulting impression gained from all the meetings was that there were a small number of progressive companies that saw the benefits of Mechatronics but that the majority who attended remained uncommitted.

It is clear that there is a need for a targetted awareness programme to present clearly to

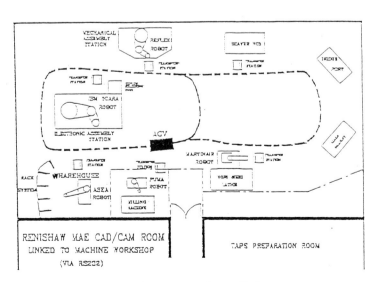

Fig 7   Mechatronics facility – Gwent College

industrialists the benefits that can be derived from adopting the Mechatronic approach to manufacture.

From the colleges' point of view it is necessary to express a concern regarding the availability of the necessary resources to enable ability to deliver to match aspirations.

5.0 THE FUTURE

To secure a strong competitive manufacturing base must surely be the long term aim of the U.K. This would enable us to compete competitively in world markets and ensure that we maintain and improve on our existing standard of living. Over the past four years many SME's and in some instances large companies have ceased trading.

To build for the future we must ensure that the lessons from the past are heeded. Companies must examine the type and range of products they manufacture and analyse the alternative manufacturing systems taking account of both the supplier and consumer ends of the system.

The consumer product market is continually demanding more sophisticated, intelligent products. The "smart" solutions to design problems are in demand.

A possible scenario, therefore, for the future would be one that would require industry to produce intelligent, "smart" products, for example buildings, ships, automatic guided vehicles, traffic control systems, automobiles and aircraft etc. in the context of the manufacture of the total system.

If this "total system manufacture" philosophy was adopted it would provide greater flexibility within the design/manufacture/maintenance process and allow ideas to be changed quickly into quality products.

Such a future is not difficult to visualise. We have progressed from the gramophone to the CD player, the main frame computer to the PC note book all within a short space of time. As we move into the 21st century we could equally see similar major steps in the application of technology.

To achieve this vision of the future from our existing base of mainly traditional management and manufacturing methods will need a considerable amount of investment in both equipment and human resources. With the increase in sophistication of product comes greater demands on excellence in manufacture and a workforce that has intellectual mobility across discipline interfaces. The final stage of this development process must be the establishment of the "dark factory", the ultimate in flexibility and quality.

This must be the direction in which the manufacturing industry should be driven to retrieve our capability to compete in World markets and particularly against the Asian Ring. Unfortunately the UK has fallen so badly behind in the competitive stakes that in order to do this we need a "step function" change in our manufacturing capability.

The provision of the required capital investment must be the responsibility of Government and industrialists. The provision of the suitably educated and trained human resource to support the industry will be the responsibility of Educational Establishments which are suitably resourced in equipment and staff. Because of the high financial costs involved in continually updating machines, software and staff, centres should be nominated and receive government and industrial support.

There are, as we know, difficulties in attracting young people into engineering. This situation is likely to continue until the profession gains recognition equal to that which exists in our partner countries in Europe. However there are at present a large number of BTEC qualified and experienced people available for retraining/updating in Mechatronics. They, together with their employers, would need convincing of the benefits that can be derived from the Mechatronics approach. There is very little evidence at present that this message has been received and understood.

It was the DTI skills survey in 1984 that led eventually to the publication in May 1991 of the BTEC Higher National Certificate programme in Mechatronics. We have a short time scale available to us in which to rebuild and strengthen our manufacturing industry. The debate therefore about the role of Mechatronics in our industrial future should now be undertaken at Board Room level between educationalists and the managers.

If the programme is to proceed then there must be joint ownership to ensure that the necessary support and outcomes match the needs of both partners.

REFERENCES:

(1) Employment Gazette Vol. 99 No. 4 HMSO April 1991

(2) Internal report: Introduction of Mechatronics - BTEC November 1989.

(3) DINSDALE J. Mechatronics: The International Scene: Mechatronics Systems Engineering, Vol. 1 No. 2 November 1990.

(4) Internal "State of Play" Report. BTEC October 1990.

(5) Engineering (Mechatronics) Guidance and Units. BTEC Higher National Awards Publication No. 07-018-1. May 1991.